TAKE NOTE!

TO ACCOMPANY

FUNDAMENTALS OF BIOCHEMISTRY

Donald Voet
University of Pennsylvania

Judith G. Voet
Swarthmore College

Charlotte W. Pratt
Seattle, Washington

Prepared by
Laura Ierardi
LCI Design

JOHN WILEY & SONS, INC.
New York • Chichester • Weinheim
Brisbane • Singapore • Toronto

The front cover shows some of the molecular assemblies that form the circle of life: *DNA makes RNA makes protein makes DNA.*

The images are (*clockwise from the top*):

1. B-DNA, *based on an X-ray structure by Richard Dickerson and Horace Drew.*
2. The nucleosome, *courtesy of Timothy Richmond.*
3. Model of the *lac* repressor in complex with DNA and CAP protein, *courtesy of Ponzy Lu and Mitchell Lewis.*
4. Ribozyme RNA, *based on an X-ray structure by Jennifer Doudna.*
5. The ribosome in complex with tRNAs, *courtesy of Joachim Frank.*
6. DNA polymerase in complex with DNA, *courtesy of Tom Ellenberger.*

The central image is based on Leonardo da Vinci's drawing *Study of Proportions.* It represents for us the never ending human quest for understanding. (© G. Bartholomew/ Westlight)

ISBN 0-471-33049-3

Printed in the United States of America

10 9 8 7 6 5 4 3 2

Printed and bound by Courier Westford, Inc.

Table of Contents

LIFE

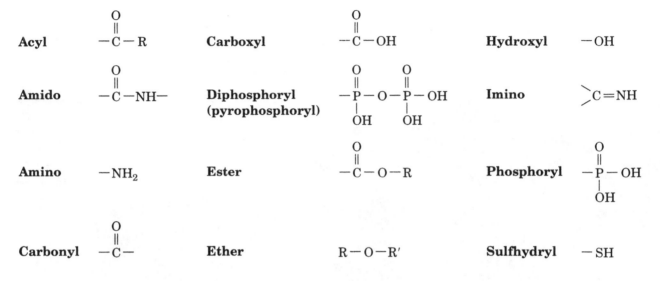

Figure 1-2.   *Key to Structure.* Common functional groups in biochemistry.

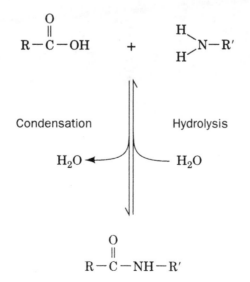

Figure 1-3.   Reaction of a carboxylic acid with an amine.

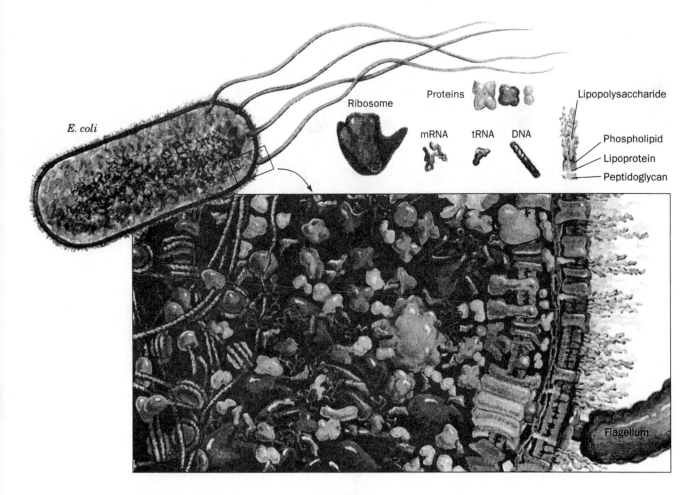

Labels in figure: E. coli, Ribosome, Proteins, mRNA, tRNA, DNA, Lipopolysaccharide, Phospholipid, Lipoprotein, Peptidoglycan, Flagellum

Figure 1-6. Cross section of an *E. coli* cell.

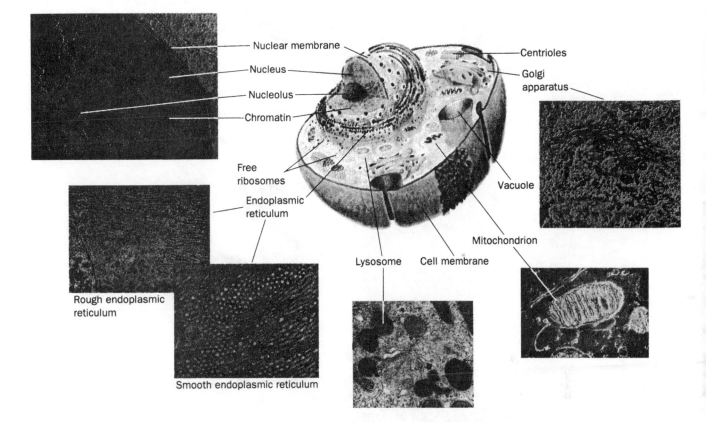

Figure 1-8. **Diagram of a typical animal cell accompanied by electron micrographs of its organelles.**

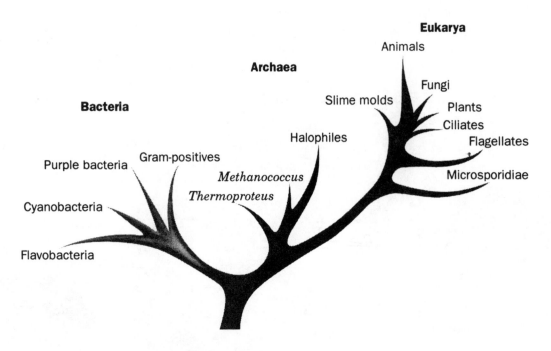

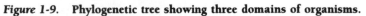

Figure 1-9. Phylogenetic tree showing three domains of organisms.

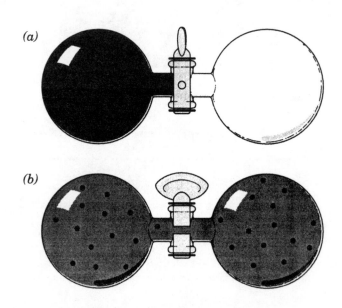

Figure 1-10. Random distribution of gas molecules.

$$S = k_B \ln W \qquad\qquad [1\text{-}5]$$

$$\Delta S_{\text{system}} + \Delta S_{\text{surroundings}} = \Delta S_{\text{universe}} > 0 \qquad [1\text{-}6]$$

$$\Delta S \geq \frac{q}{T} \qquad\qquad [1\text{-}7]$$

$$\Delta S \geq \frac{q_P}{T} = \frac{\Delta H}{T} \qquad\qquad [1\text{-}8]$$

$$\Delta H - T\Delta S \leq 0 \qquad\qquad [1\text{-}9]$$

$$G = H - TS \qquad\qquad [1\text{-}10]$$

$$\boxed{\Delta G = \Delta H - T\Delta S < 0} \qquad\qquad [1\text{-}11]$$

Table 1-3. **Variation of Reaction Spontaneity (Sign of ΔG) with the Signs of ΔH and ΔS**

ΔH	ΔS	$\Delta G = \Delta H - T\Delta S$
−	+	The reaction is both enthalpically favored (exothermic) and entropically favored. It is spontaneous (exergonic) at all temperatures.
−	−	The reaction is enthalpically favored but entropically opposed. It is spontaneous only at temperatures *below* $T = \Delta H/\Delta S$.
+	+	The reaction is enthalpically opposed (endothermic) but entropically favored. It is spontaneous only at temperatures *above* $T = \Delta H/\Delta S$.
+	−	The reaction is both enthalpically and entropically opposed. It is *un*spontaneous (endergonic) at all temperatures.

$$\overline{G}_A - \overline{G}_A^\circ = RT \ln [A] \qquad\qquad [1\text{-}12]$$

$$aA + bB \rightleftharpoons cC + dD$$

$$\Delta G = c\overline{G}_C + d\overline{G}_D - a\overline{G}_A - b\overline{G}_B \qquad\qquad [1\text{-}13]$$

$$\Delta G^\circ = c\overline{G}_C^\circ + d\overline{G}_D^\circ - a\overline{G}_A^\circ - b\overline{G}_B^\circ \qquad\qquad [1\text{-}14]$$

$$\Delta G = \Delta G^\circ + RT \ln \left(\frac{[C]^c [D]^d}{[A]^a [B]^b} \right) \qquad\qquad [1\text{-}15]$$

$$\boxed{\Delta G^\circ = -RT \ln K_{eq}} \qquad\qquad [1\text{-}16]$$

$$K_{eq} = \frac{[C]_{eq}^c [D]_{eq}^d}{[A]_{eq}^a [B]_{eq}^b} = e^{-\Delta G^\circ / RT} \qquad\qquad [1\text{-}17]$$

$$\ln K_{eq} = \frac{-\Delta H^\circ}{R} \left(\frac{1}{T} \right) + \frac{\Delta S^\circ}{R} \qquad\qquad [1\text{-}18]$$

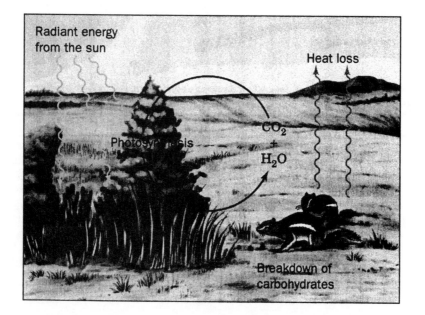

Figure 1-11. **Energy flow in the biosphere.**

CHAPTER 2

WATER

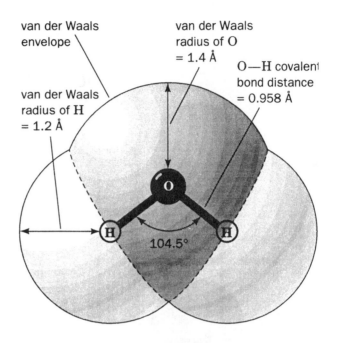

Figure 2-1. Structure of the water molecule.

Table 2-1. Bond Energies in Biomolecules

Type of Bond	Example	Bond Strength $(kJ \cdot mol^{-1})$
Covalent	O—H	460
	C—H	414
	C—C	348
Noncovalent		
Ionic interaction	—COO$^-$----$^+$H$_3$N—	86
Hydrogen bond	—O—H---O$\diagdown$	20
van der Waals forces		
Dipole–dipole interaction	$\diagdown$C=O---$\diagdown$C=O	9.3
London dispersion force	—C—H---H—C—	0.3

(a) Interactions between permanent dipoles

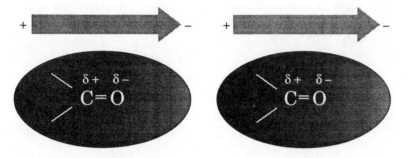

(b) Dipole–induced dipole interactions

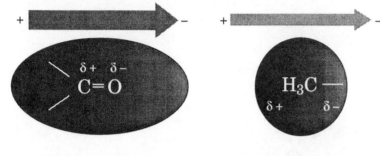

(c) London dispersion forces

 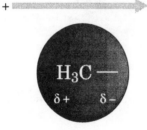

Figure 2-5. Dipole–dipole interactions.

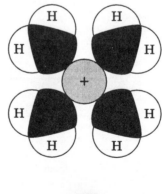

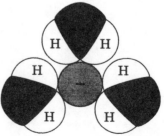

Figure 2-6. Solvation of ions.

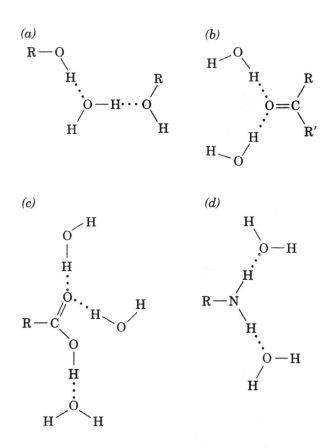

Figure 2-7. Hydrogen bonding by functional groups.

Table 2-2. Thermodynamic Changes for Transferring Hydrocarbons from Water to Nonpolar Solvents at 25°C

Process	ΔH $(kJ \cdot mol^{-1})$	$-T\Delta S$ $(kJ \cdot mol^{-1})$	ΔG $(kJ \cdot mol^{-1})$
CH_4 in $H_2O \rightleftharpoons CH_4$ in C_6H_6	11.7	−22.6	−10.9
CH_4 in $H_2O \rightleftharpoons CH_4$ in CCl_4	10.5	−22.6	−12.1
C_2H_6 in $H_2O \rightleftharpoons C_2H_6$ in benzene	9.2	−25.1	−15.9
C_2H_4 in $H_2O \rightleftharpoons C_2H_4$ in benzene	6.7	−18.8	−12.1
C_2H_2 in $H_2O \rightleftharpoons C_2H_2$ in benzene	0.8	−8.8	−8.0
Benzene in $H_2O \rightleftharpoons$ liquid benzene[a]	0.0	−17.2	−17.2
Toluene in $H_2O \rightleftharpoons$ liquid toluene[a]	0.0	−20.0	−20.0

[a]Data measured at 18°C.

Source: Kauzmann, W., *Adv. Protein Chem.* **14,** 39 (1959).

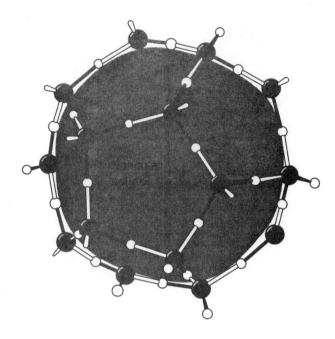

Figure 2-8. Orientation of water molecules around a nonpolar solute.

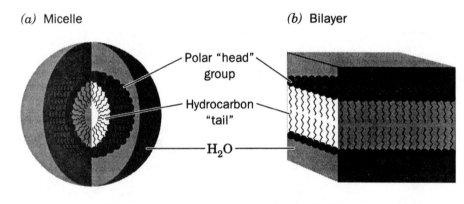

(a) Micelle

(b) Bilayer

Polar "head" group

Hydrocarbon "tail"

H_2O

Figure 2-10. Structures of micelles and bilayers.

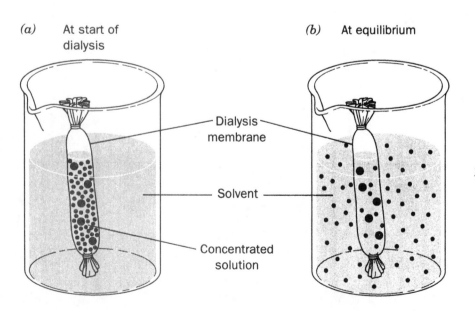

(a) At start of dialysis

(b) At equilibrium

Dialysis membrane

Solvent

Concentrated solution

Figure 2-12. Dialysis.

$$K_w = [\text{H}^+][\text{OH}^-] \qquad [2\text{-}2]$$

$$\text{pH} = -\log [\text{H}^+] \qquad [2\text{-}3]$$

Table 2-4. pH Values of Some Common Substances

Substance	pH
1 M NaOH	14
Household ammonia	12
Seawater	8
Blood	7.4
Milk	7
Saliva	6.6
Tomato juice	4.4
Vinegar	3
Gastric juice	1.5
1 M HCl	0

$$K = \frac{[\text{H}_3\text{O}^+]\,[\text{A}^-]}{[\text{HA}]\,[\text{H}_2\text{O}]} \qquad [2\text{-}4]$$

$$K_a = K[\text{H}_2\text{O}] = \frac{[\text{H}^+][\text{A}^-]}{[\text{HA}]} \qquad [2\text{-}5]$$

$$\text{p}K = -\log K \qquad [2\text{-}6]$$

$$[\text{H}^+] = K\frac{[\text{HA}]}{[\text{A}^-]} \qquad [2\text{-}7]$$

$$\text{pH} = -\log K + \log \frac{[\text{A}^-]}{[\text{HA}]} \qquad [2\text{-}8]$$

$$\boxed{\text{pH} = \text{p}K + \log \frac{[\text{A}^-]}{[\text{HA}]}} \qquad [2\text{-}9]$$

Table 2-5. Dissociation Constants and pK Values at 25°C of Some Acids

Acid	K	pK
Oxalic acid	5.37×10^{-2}	1.27 (pK_1)
H_3PO_4	7.08×10^{-3}	2.15 (pK_1)
Formic acid	1.78×10^{-4}	3.75
Succinic acid	6.17×10^{-5}	4.21 (pK_1)
Oxalate$^-$	5.37×10^{-5}	4.27 (pK_2)
Acetic acid	1.74×10^{-5}	4.76
Succinate$^-$	2.29×10^{-6}	5.64 (pK_2)
2-(N-Morpholino)ethanesulfonic acid (MES)	8.13×10^{-7}	6.09
H_2CO_3	4.47×10^{-7}	6.35 (pK_1)
Piperazine-N,N'-bis(2-ethanesulfonic acid) (PIPES)	1.74×10^{-7}	6.76
$H_2PO_4^-$	1.51×10^{-7}	6.82 (pK_2)
3-(N-Morpholino)propanesulfonic acid (MOPS)	7.08×10^{-8}	7.15
N-2-Hydroxyethylpiperazine-N'-2-ethanesulfonic acid (HEPES)	3.39×10^{-8}	7.47
Tris(hydroxymethyl)aminomethane (Tris)	8.32×10^{-9}	8.08
NH_4^+	5.62×10^{-10}	9.25
Glycine	1.66×10^{-10}	9.78
HCO_3^-	4.68×10^{-11}	10.33 (pK_2)
Piperidine	7.58×10^{-12}	11.12
HPO_4^{2-}	4.17×10^{-13}	12.38 (pK_3)

Source: Dawson, R.M.C., Elliott, D.C., Elliott, W.H., and Jones, K.M., **Data for Biochemical Research** (3rd ed.), pp. 424–425, Oxford Science Publications (1986) *and* Good, N.E., Winget, G.D., Winter, W., Connolly, T.N., Izawa, S., and Singh, R.M.M., *Biochemistry* **5**, 467 (1966).

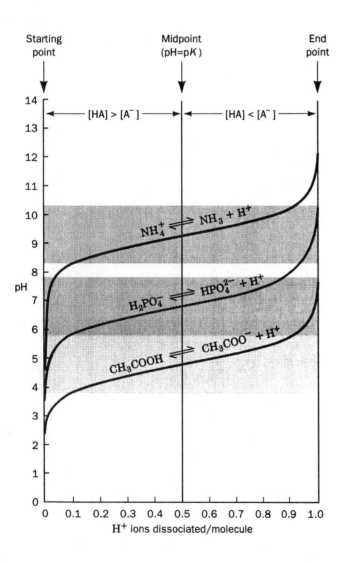

Figure 2-15. **Titration curves for acetic acid, phosphate, and ammonia.**

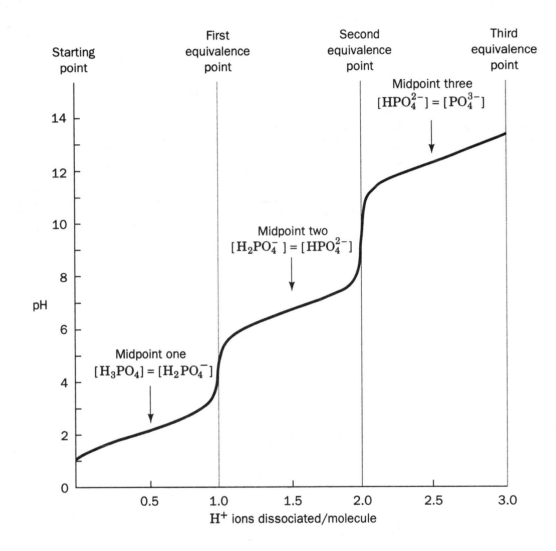

Figure 2-16. **Titration of a polyprotic acid.**

CHAPTER 3

NUCLEOTIDES AND NUCLEIC ACIDS

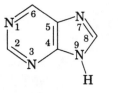

Purine   **Pyrimidine**

Table 3-1. **Names and Abbreviations of Nucleic Acid Bases, Nucleosides, and Nucleotides**

Base Formula	Base (X = H)	Nucleoside (X = ribose[a])	Nucleotide[b] (X = ribose phosphate[a])
Adenine structure	Adenine / Ade / A	Adenosine / Ado / A	Adenylic acid / Adenosine monophosphate / AMP
Guanine structure	Guanine / Gua / G	Guanosine / Guo / G	Guanylic acid / Guanosine monophosphate / GMP
Cytosine structure	Cytosine / Cyt / C	Cytidine / Cyd / C	Cytidylic acid / Cytidine monophosphate / CMP
Uracil structure	Uracil / Ura / U	Uridine / Urd / U	Uridylic acid / Uridine monophosphate / UMP
Thymine structure	Thymine / Thy / T	Deoxythymidine / dThd / dT	Deoxythymidylic acid / Deoxythymidine monophosphate / dTMP

[a]The presence of a 2′-deoxyribose unit in place of ribose, as occurs in DNA, is implied by the prefixes "deoxy" or "d." For example, the deoxynucleoside of adenine is deoxyadenosine or dA. However, for thymine-containing residues, which rarely occur in RNA, the prefix is redundant and may be dropped. The presence of a ribose unit may be explicitly implied by the prefix "ribo" or "r." Thus the ribonucleotide of thymine is ribothymidine or rT.

[b]The position of the phosphate group in a nucleotide may be explicitly specified as in, for example, 3′-AMP and 5′-GMP.

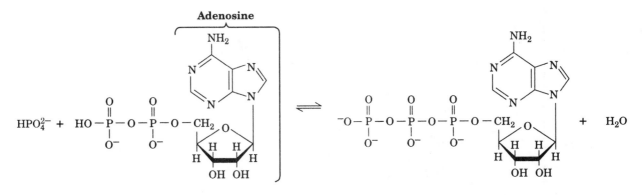

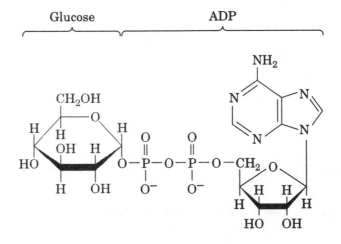

Figure 3-2. **ADP–glucose.**

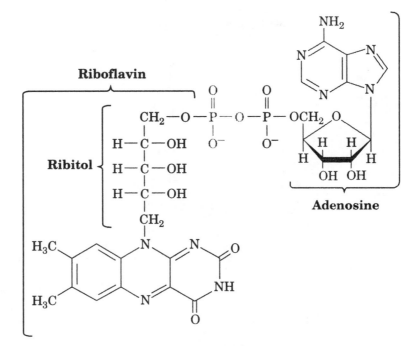

Figure 3-3. **Flavin adenine dinucleotide (FAD).**

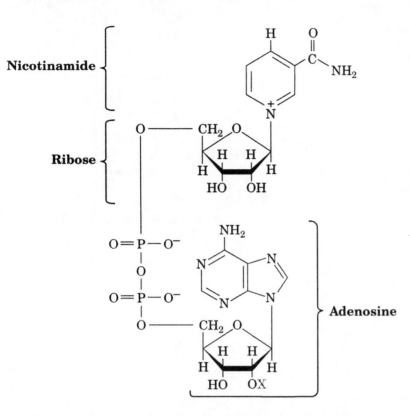

X = H   **Nicotinamide adenine dinucleotide (NAD$^+$)**
X = PO$_3^{2-}$ **Nicotinamide adenine dinucleotide phosphate (NADP$^+$)**

Figure 3-4.   Nicotinamide adenine dinucleotide (NAD$^+$) and nicotinamide adenine dinucleotide phosphate (NADP$^+$).

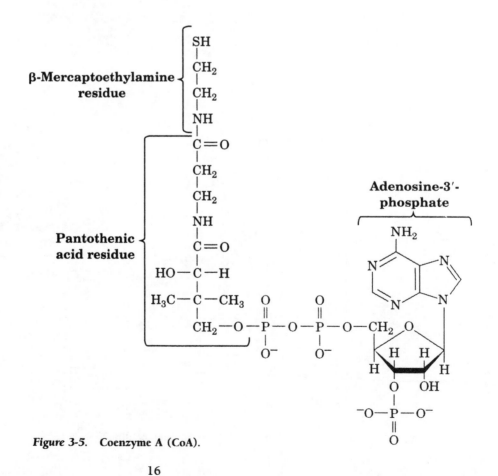

Figure 3-5.   Coenzyme A (CoA).

16

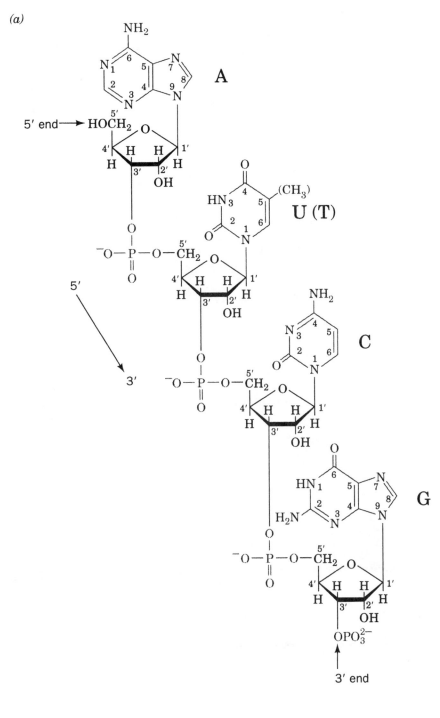

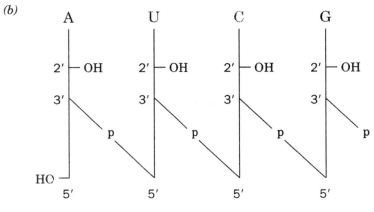

Figure 3-6. Key to Structure. Chemical structure of a nucleic acid.

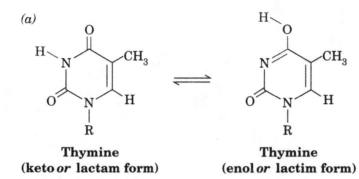

(a)

Thymine
(keto *or* lactam form)

Thymine
(enol *or* lactim form)

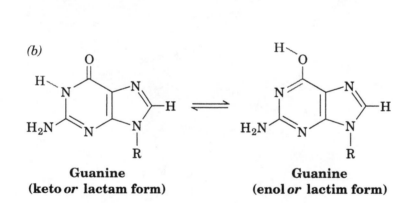

(b)

Guanine
(keto *or* lactam form)

Guanine
(enol *or* lactim form)

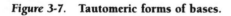

Figure 3-7. **Tautomeric forms of bases.**

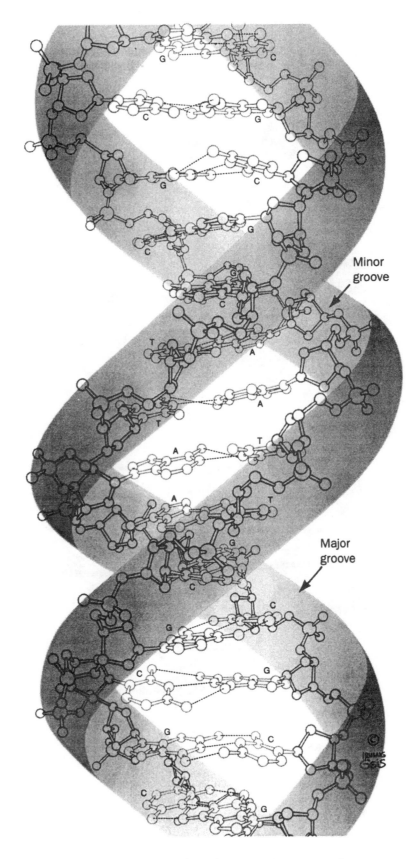

Minor groove

Major groove

Figure 3-9. **Three-dimensional structure of DNA.**

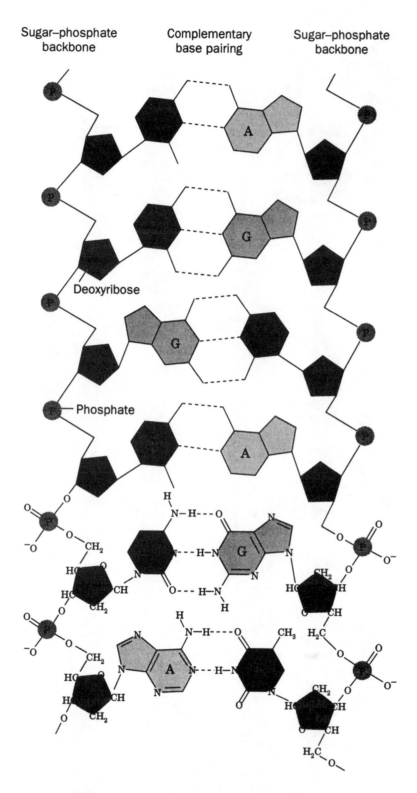

Figure 3-11. Complementary strands of DNA.

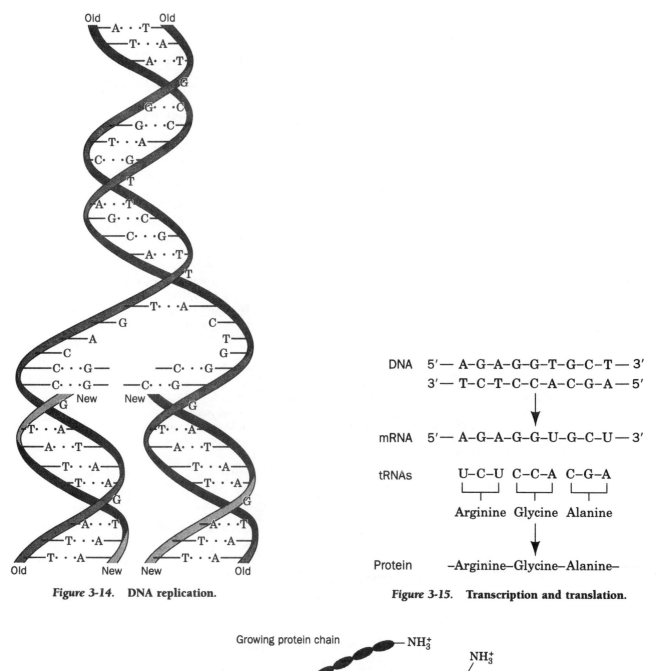

Figure 3-14. DNA replication.

DNA   5′ — A–G–A–G–G–T–G–C–T — 3′
      3′ — T–C–T–C–C–A–C–G–A — 5′

mRNA 5′ — A–G–A–G–G–U–G–C–U — 3′

tRNAs   U–C–U C–C–A C–G–A

          Arginine Glycine Alanine

Protein   –Arginine–Glycine–Alanine–

Figure 3-15.   Transcription and translation.

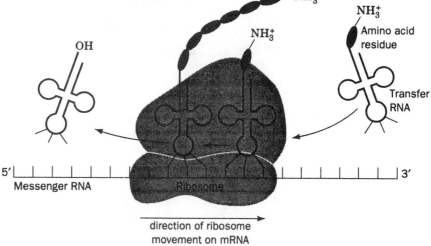

Growing protein chain

Figure 3-16.   Translation.

Table 3-3. Recognition and Cleavage Sites of Some Type II Restriction Enzymes

Enzyme	Recognition Sequence[a]	Microorganism
AluI	AG↓C*T	Arthrobacter luteus
BamHI	G↓GATC*C	Bacillus amyloliquefaciens H
BglI	GCCNNNN↓NGGC	Bacillus globigii
BglII	A↓GATCT	Bacillus globigii
EcoRI	G↓AA*TTC	Escherichia coli RY13
EcoRII	↓CC*(A̶)GG	Escherichia coli R245
EcoRV	GA*T↓ATC	Escherichia coli J62pLG74
HaeII	RGCGC↓Y	Haemophilus aegyptius
HaeIII	GG↓C*C	Haemophilus aegyptius
HindIII	A*↓AGCTT	Haemophilus influenzae R$_d$
HpaII	C↓C*GG	Haemophilus parainfluenzae
MspI	C*↓CGG	Moraxella species
PstI	CTGCA*↓G	Providencia stuartii 164
PvuII	CAG↓C*TG	Proteus vulgaris
SalI	G↓TCGAC	Streptomyces albus G
TaqI	T↓CGA*	Thermus aquaticus
XhoI	C↓TCGAG	Xanthomonas holcicola

[a]The recognition sequence is abbreviated so that only one strand, reading 5′ to 3′, is given. The cleavage site is represented by an arrow (↓) and the modified base, where it is known, is indicated by an asterisk (A* is N^6-methyladenine and C* is 5-methylcytosine). R, Y, and N represent purine nucleotide, pyrimidine nucleotide, and any nucleotide, respectively.

Source: Roberts, R.J. and Macellis, D., REBASE—the restriction enzyme database, http://www.neb.com/rebase.

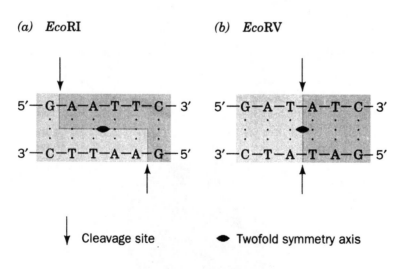

(a) EcoRI   (b) EcoRV

↓ Cleavage site   ● Twofold symmetry axis

Figure 3-18. Restriction sites.

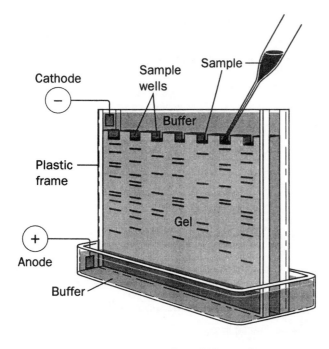

Figure 3-19. Apparatus for gel electrophoresis.

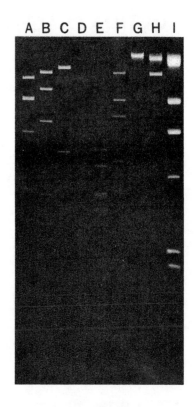

Figure 3-20. Electrophoretogram of restriction digests.

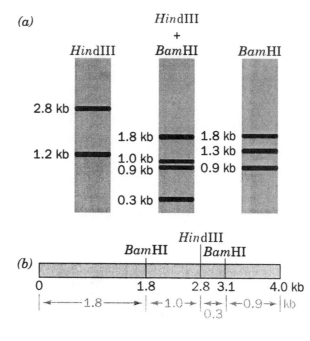

Figure 3-21. Construction of a restriction map.

RESTRICTION FRAGMENT LENGTH POLYMORPHISMS (RFLPs)

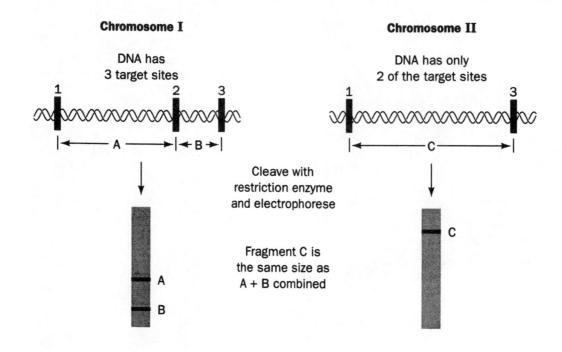

Chromosome I

DNA has
3 target sites

Chromosome II

DNA has only
2 of the target sites

Cleave with
restriction enzyme
and electrophorese

Fragment C is
the same size as
A + B combined

THE CHAIN-TERMINATOR METHOD OF SEQUENCING DNA

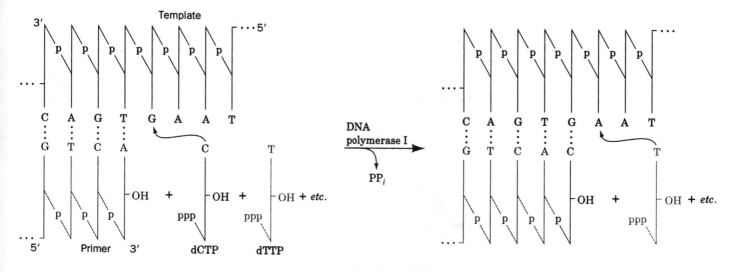

Figure 3-22. Action of DNA polymerase I.

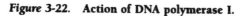

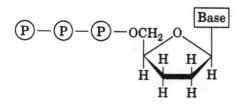

**2′,3′-Dideoxynucleoside
triphosphate**

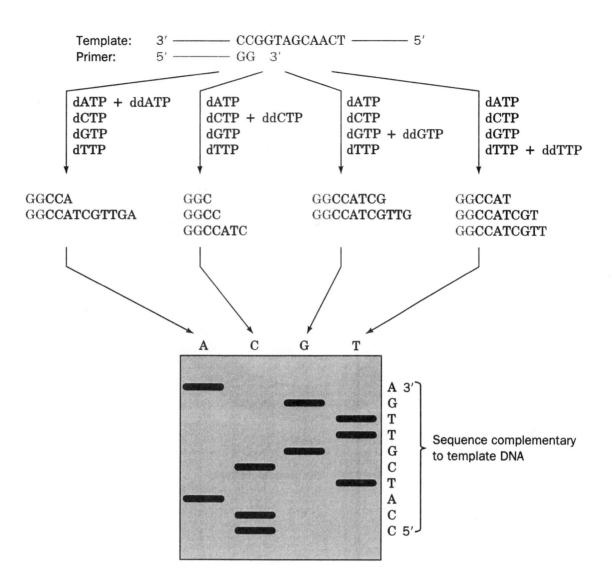

Figure 3-23. The chain-terminator (dideoxy) method of DNA sequencing.

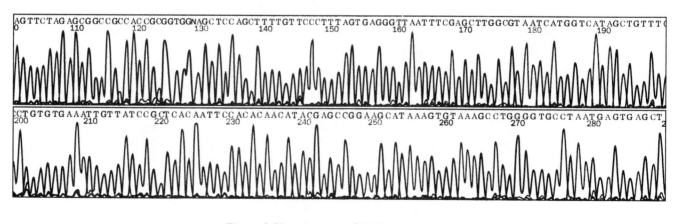

Figure 3-25. Automated DNA sequencing.

25

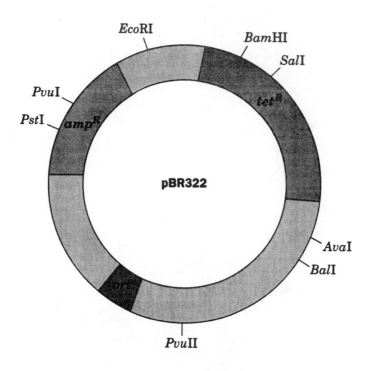

Figure 3-27. The plasmid pBR322.

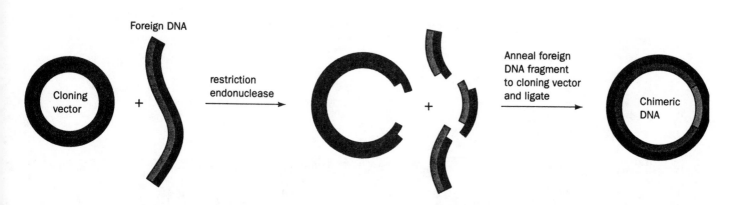

Figure 3-29. Construction of a recombinant DNA molecule.

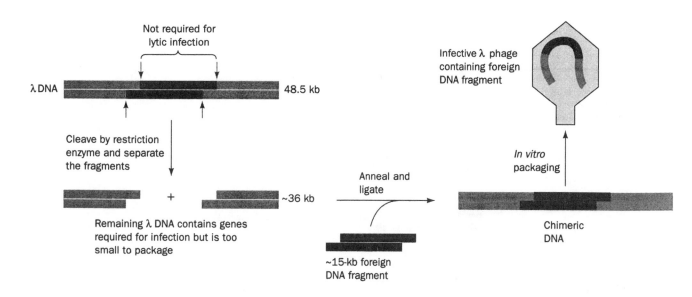

Not required for lytic infection

λ DNA — 48.5 kb

Cleave by restriction enzyme and separate the fragments

+ ~36 kb

Remaining λ DNA contains genes required for infection but is too small to package

Anneal and ligate

~15-kb foreign DNA fragment

Infective λ phage containing foreign DNA fragment

In vitro packaging

Chimeric DNA

***Figure 3-30.* Cloning with bacteriophage λ.**

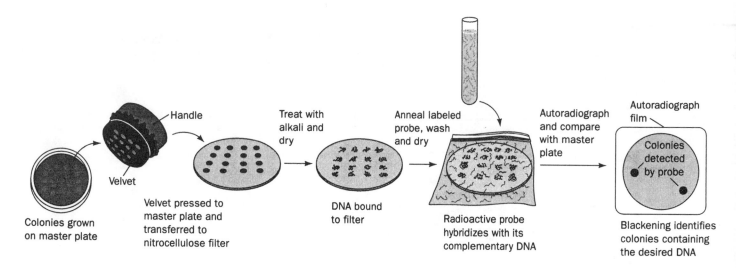

Colonies grown on master plate

Handle
Velvet

Velvet pressed to master plate and transferred to nitrocellulose filter

Treat with alkali and dry

DNA bound to filter

Anneal labeled probe, wash and dry

Radioactive probe hybridizes with its complementary DNA

Autoradiograph and compare with master plate

Autoradiograph film

Colonies detected by probe

Blackening identifies colonies containing the desired DNA

***Figure 3-31.* Colony (*in situ*) hybridization.**

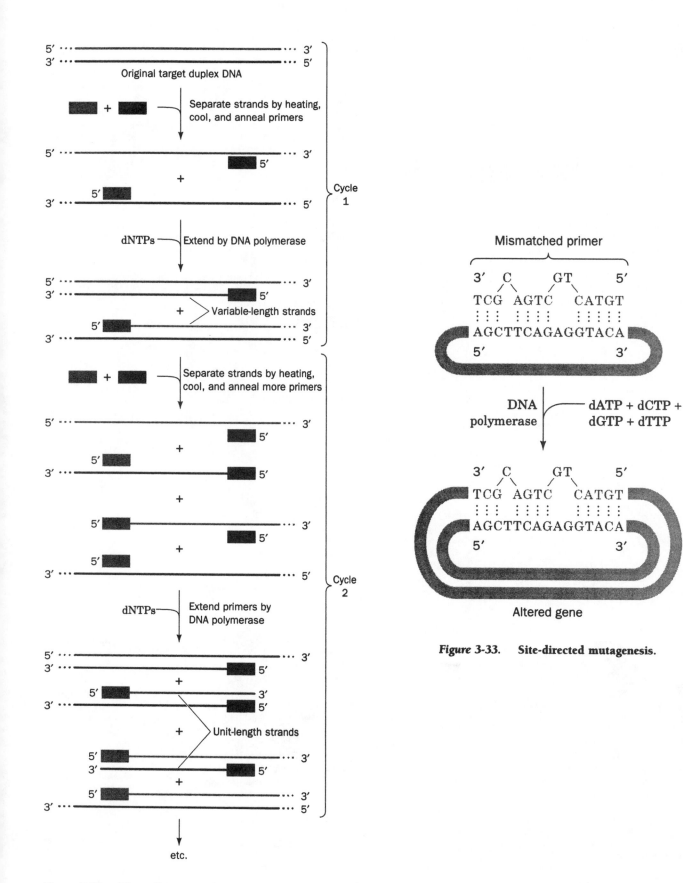

Figure 3-33. Site-directed mutagenesis.

Figure 3-32. The polymerase chain reaction (PCR).

AMINO ACIDS

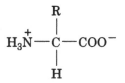

Figure 4-2.   A zwitterionic amino acid.

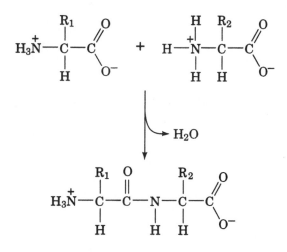

Figure 4-3.   Condensation of two amino acids.

Table 4-1. Key to Structure. Covalent Structures and Abbreviations of the "Standard" Amino Acids of Proteins, Their Occurrence, and the pK Values of Their Ionizable Groups

Name, Three-letter Symbol, and One-letter Symbol	Structural Formula[a]	Residue Mass (D)[b]	Average Occurrence in Proteins (%)[c]	pK_1 α-COOH[d]	pK_2 α-NH$_3^+$[d]	pK_R Side Chain[d]
Amino acids with nonpolar side chains						
Glycine Gly G	COO⁻ H—C—H NH₃⁺	57.0	7.2	2.35	9.78	
Alanine Ala A	COO⁻ H—C—CH₃ NH₃⁺	71.1	7.8	2.35	9.87	
Valine Val V	COO⁻ H—C—CH(CH₃)₂ NH₃⁺	99.1	6.6	2.29	9.74	
Leucine Leu L	COO⁻ H—C—CH₂—CH(CH₃)₂ NH₃⁺	113.2	9.1	2.33	9.74	
Isoleucine Ile I	COO⁻ H—C—C*—CH₂—CH₃ NH₃⁺	113.2	5.3	2.32	9.76	
Methionine Met M	COO⁻ H—C—CH₂—CH₂—S—CH₃ NH₃⁺	131.2	2.2	2.13	9.28	
Proline Pro P	COO⁻ C²—CH₂ (ring)	97.1	5.2	1.95	10.64	
Phenylalanine Phe F	COO⁻ H—C—CH₂—⬡ NH₃⁺	147.2	3.9	2.20	9.31	
Tryptophan Trp W	COO⁻ H—C—CH₂—(indole) NH₃⁺	186.2	1.4	2.46	9.41	

[a]The ionic forms shown are those predominating at pH 7.0 although residue mass is given for the neutral compound The C_α atoms, as well as those atoms marked with an asterisk, are chiral centers with configurations as indicated according to Fischer projection formulas. The standard organic numbering system is provided for heterocycles.

[b]The residue masses are given for the neutral residues. For the molecular masses of the parent amino acids, add 18.0 D, the molecular mass of H_2O, to the residue masses. For side chain masses, subtract 56.0 D, the formula mass of a peptide group, from the residue masses.

[c]Calculated from a database of nonredundant proteins containing 300,688 residues as compiled by Doolittle, R.F. in Fasman, G.D. (Ed.), *Predictions of Protein Structure and the Principles of Protein Conformation,* Plenum Press (1989).

[d]Data from Dawson, R.M.C., Elliott, D.C., Elliott, W.H., and Jones, K.M., *Data for Biochemical Research* (3rd ed.), pp. 1–31, Oxford Science Publications (1986).

[e]The three- and one-letter symbols for asparagine *or* aspartic acid are Asx and B, whereas for glutamine *or* glutamic acid they are Glx and Z. The one-letter symbol for an undetermined or "nonstandard" amino acid is X.

[f]Both neutral and protonated forms of histidine are present at pH 7.0, since pK_R is close to 7.0.

Table 4-1. (continued)

Name, Three-letter Symbol, and One-letter Symbol	Structural Formula[a]	Residue Mass (D)[b]	Average Occurrence in Proteins (%)[c]	pK_1 α-COOH[d]	pK_2 α-NH_3^+[d]	pK_R Side Chain[d]
Amino acids with uncharged polar side chains						
Serine Ser S		87.1	6.8	2.19	9.21	
Threonine Thr T		101.1	5.9	2.09	9.10	
Asparagine[e] Asn N		114.1	4.3	2.14	8.72	
Glutamine[e] Gln Q		128.1	4.3	2.17	9.13	
Tyrosine Tyr Y		163.2	3.2	2.20	9.21	10.46 (phenol)
Cysteine Cys C		103.1	1.9	1.92	10.70	8.37 (sulfhydryl)
Amino acids with charged polar side chains						
Lysine Lys K		128.2	5.9	2.16	9.06	10.54 (ϵ-NH_3^+)
Arginine Arg R		156.2	5.1	1.82	8.99	12.48 (guanidino)
Histidine[f] His H		137.1	2.3	1.80	9.33	6.04 (imidazole)
Aspartic acid[e] Asp D		115.1	5.3	1.99	9.90	3.90 (β-COOH)
Glutamic acid[e] Glu E		129.1	6.3	2.10	9.47	4.07 (γ-COOH)

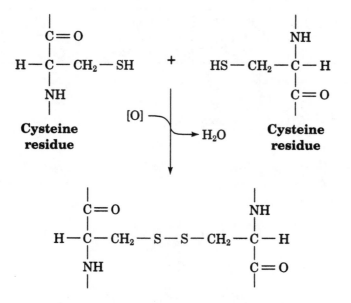

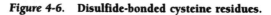

Figure 4-6. Disulfide-bonded cysteine residues.

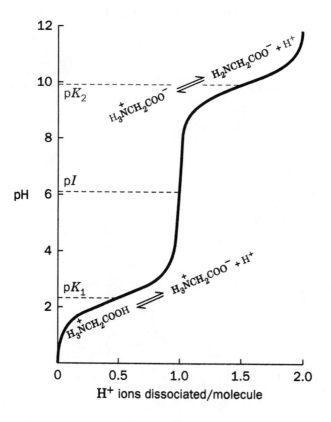

Figure 4-8. Titration of glycine.

$$pH = pK + \log \frac{[A^-]}{[HA]} \qquad [4\text{-}1]$$

$$pI = \tfrac{1}{2}(pK_i + pK_j) \qquad [4\text{-}2]$$

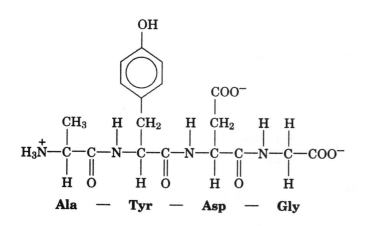

Ala — Tyr — Asp — Gly

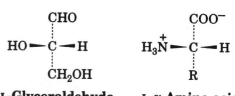

L-Glyceraldehyde   L-α-Amino acid

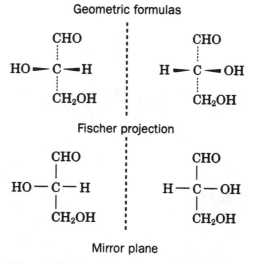

Figure 4-12. The Fischer convention.

PROTEINS: PRIMARY STRUCTURE

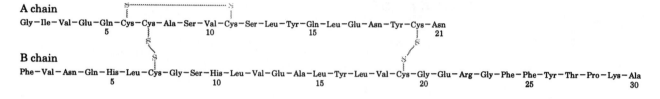

A chain

Gly—Ile—Val—Glu—Gln—Cys—Cys—Ala—Ser—Val—Cys—Ser—Leu—Tyr—Gln—Leu—Glu—Asn—Tyr—Cys—Asn
  5   10   15   21

B chain

Phe—Val—Asn—Gln—His—Leu—Cys—Gly—Ser—His—Leu—Val—Glu—Ala—Leu—Tyr—Leu—Val—Cys—Gly—Glu—Arg—Gly—Phe—Phe—Tyr—Thr—Pro—Lys—Ala
           5   10   15   20   25   30

Figure 5-1. **The primary structure of bovine insulin.**

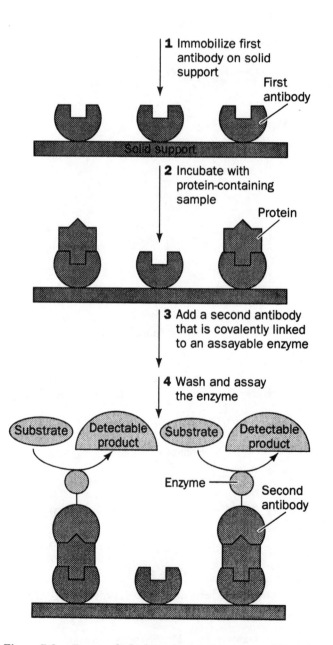

Figure 5-3. **Enzyme-linked immunosorbent assay (ELISA).**

SEPARATION TECHNIQUES

Characteristic	Procedure
Charge	Ion exchange chromatography
	Electrophoresis
Polarity	Hydrophobic interaction chromatography
Size	Gel filtration chromatography
	SDS-PAGE
	Ultracentrifugation
Binding specificity	Affinity chromatography

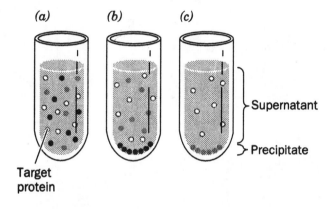

Figure 5-4. **Fractionation by salting out.**

Table 5-2. **Isoelectric Points of Several Common Proteins**

Protein	pI
Pepsin	<1.0
Ovalbumin (hen)	4.6
Serum albumin (human)	4.9
Tropomyosin	5.1
Insulin (bovine)	5.4
Fibrinogen (human)	5.8
γ-Globulin (human)	6.6
Collagen	6.6
Myoglobin (horse)	7.0
Hemoglobin (human)	7.1
Ribonuclease A (bovine)	9.4
Cytochrome c (horse)	10.6
Histone (bovine)	10.8
Lysozyme (hen)	11.0
Salmine (salmon)	12.1

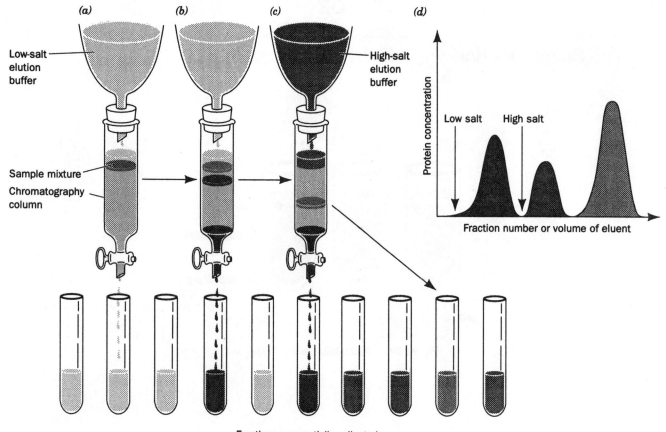

(a)

(b)

(c)

(d)

Low-salt
elution
buffer

High-salt
elution
buffer

Sample mixture

Chromatography
column

Protein concentration

Low salt   High salt

Fraction number or volume of eluent

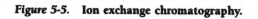

Fractions sequentially collected

Figure 5-5. **Ion exchange chromatography.**

36

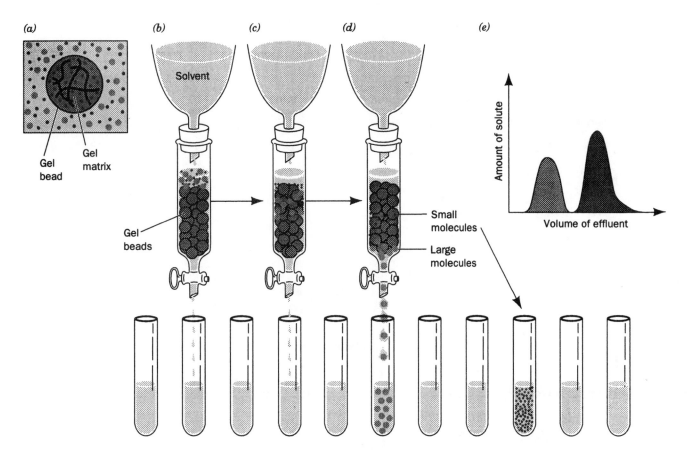

(a)

Gel
bead

Gel
matrix

(b)

Solvent

Gel
beads

(c)

(d)

Small
molecules

Large
molecules

(e)

Amount of solute

Volume of effluent

Figure 5-6. **Gel filtration chromatography.**

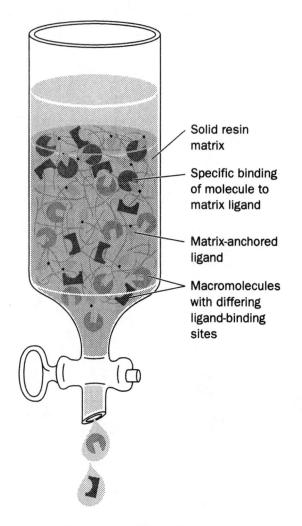

Figure 5-7. Affinity chromatography.

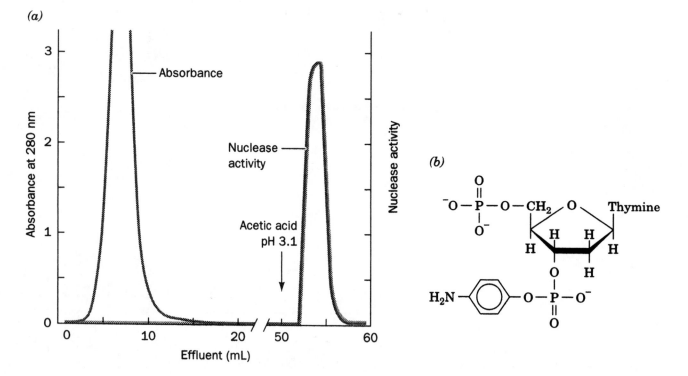

Figure 5-8. Purification of an enzyme by affinity chromatography.

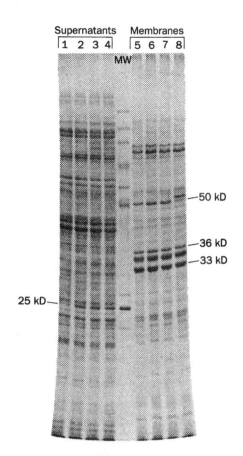

Figure 5-10. SDS-PAGE.

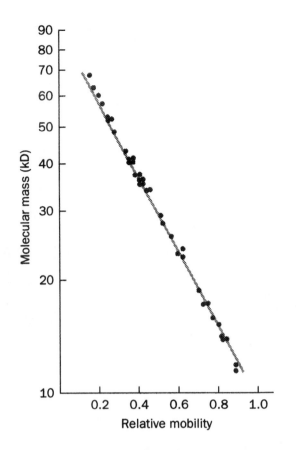

Figure 5-11. Logarithmic relationship between the molecular mass of a protein and its electrophoretic mobility in SDS-PAGE.

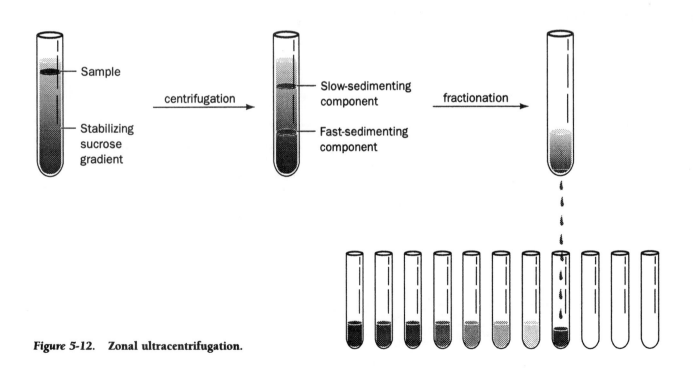

Figure 5-12. Zonal ultracentrifugation.

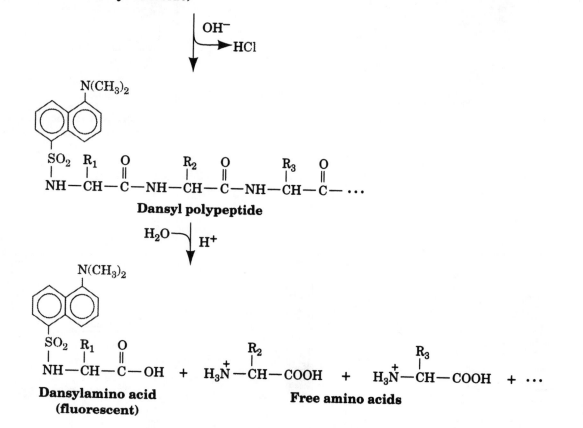

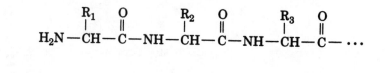

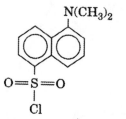

**1-Dimethylaminonaphthalene-
5-sulfonyl chloride (dansyl chloride)**

Polypeptide

Dansyl polypeptide

**Dansylamino acid
(fluorescent)**

Free amino acids

Figure 5-13. The dansyl chloride reaction.

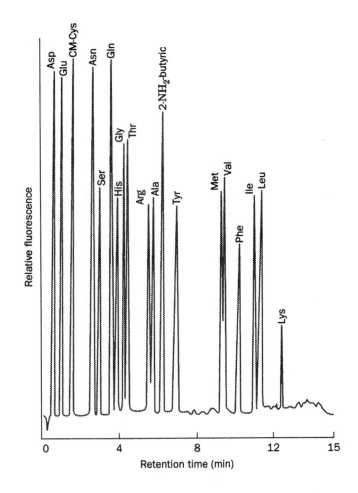

Figure 5-14. **Amino acid analysis.**

Table 5-5. **Specificities of Various Endopeptidases**

$$-NH-\underset{\underset{R_{n-1}}{|}}{CH}-\underset{\underset{O}{\|}}{C}\longrightarrow NH-\underset{\underset{R_n}{|}}{CH}-\underset{\underset{O}{\|}}{C}-$$

Scissile
peptide bond

Enzyme	Source	Specificity	Comments
Trypsin	Bovine pancreas	R_{n-1} = positively charged residues: Arg, Lys; $R_n \neq$ Pro	Highly specific
Chymotrypsin	Bovine pancreas	R_{n-1} = bulky hydrophobic residues: Phe, Trp, Tyr; $R_n \neq$ Pro	Cleaves more slowly for R_{n-1} = Asn, His, Met, Leu
Elastase	Bovine pancreas	R_{n-1} = small neutral residues: Ala, Gly, Ser, Val; $R_n \neq$ Pro	
Thermolysin	*Bacillus thermoproteolyticus*	R_n = Ile, Met, Phe, Trp, Tyr, Val; $R_{n-1} \neq$ Pro	Occasionally cleaves at R_n = Ala, Asp, His, Thr; heat stable
Pepsin	Bovine gastric mucosa	R_n = Leu, Phe, Trp, Tyr; $R_{n-1} \neq$ Pro	Also others; quite nonspecific; pH optimum = 2
Endopeptidase V8	*Staphylococcus aureus*	R_{n-1} = Glu	

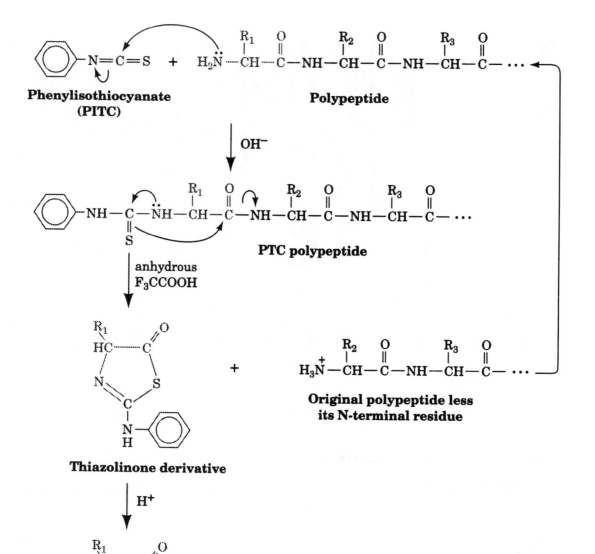

Phenylisothiocyanate (PITC)

Polypeptide

PTC polypeptide

Thiazolinone derivative

Original polypeptide less its N-terminal residue

PTH-amino acid

Figure 5-15. **Edman degradation.**

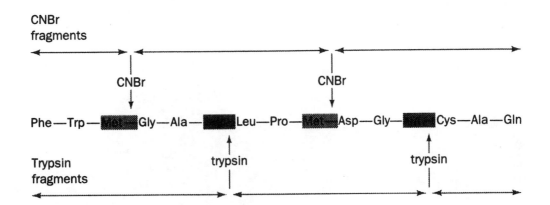

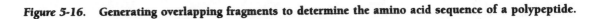

Phe—Trp—███Gly—Ala—███████Leu—Pro—███Asp—Gly—███Cys—Ala—Gln

Figure 5-16. Generating overlapping fragments to determine the amino acid sequence of a polypeptide.

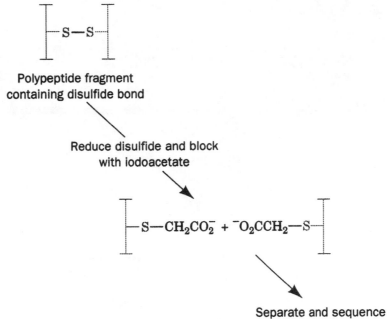

Figure 5-17. Determining the positions of disulfide bonds.

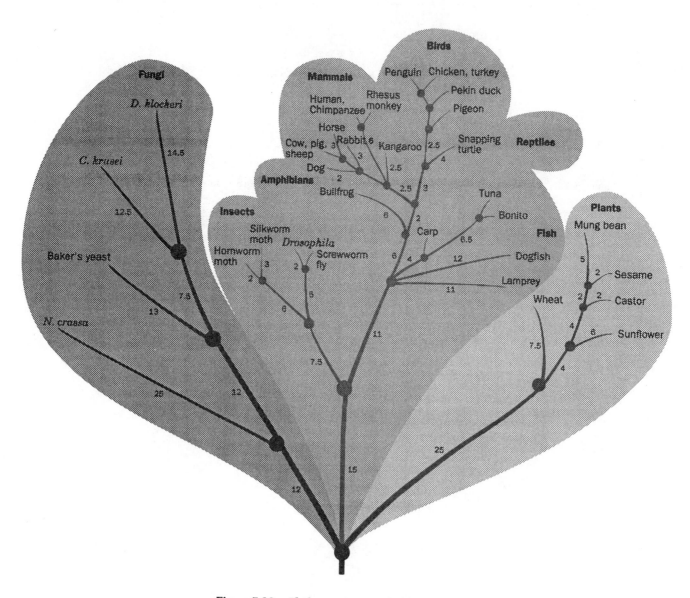

Figure 5-18. Phylogenetic tree of cytochrome *c*.

CHAPTER 6

PROTEINS: THREE-DIMENSIONAL STRUCTURE

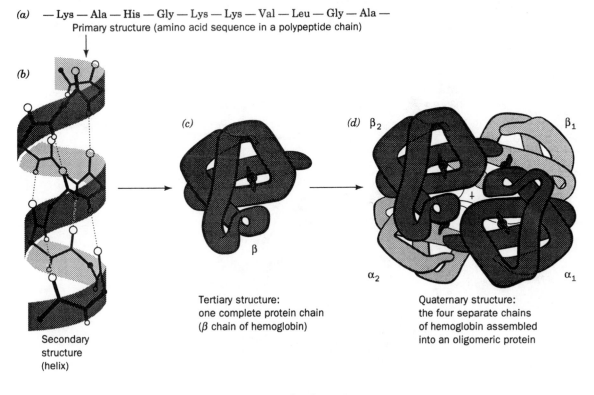

(a) — Lys — Ala — His — Gly — Lys — Lys — Val — Leu — Gly — Ala —
Primary structure (amino acid sequence in a polypeptide chain)

(b)

Secondary
structure
(helix)

(c)

Tertiary structure:
one complete protein chain
(β chain of hemoglobin)

(d) β₂ β₁

α₂ α₁

Quaternary structure:
the four separate chains
of hemoglobin assembled
into an oligomeric protein

Figure 6-1. Levels of protein structure.

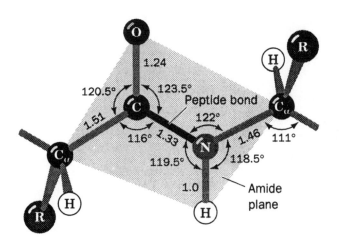

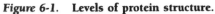

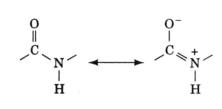

Figure 6-2. The trans peptide group.

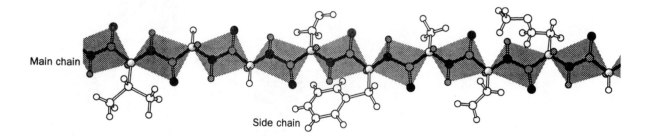

Figure 6-3. Extended conformation of a polypeptide.

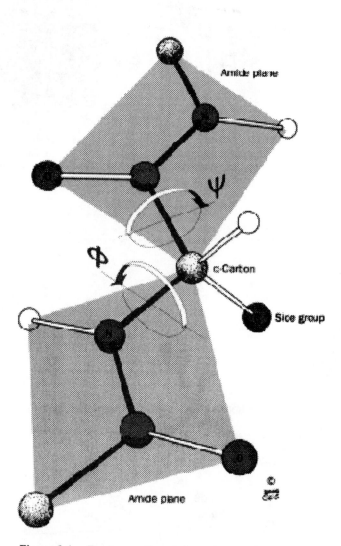

Figure 6-4. Torsion angles of the polypeptide backbone.

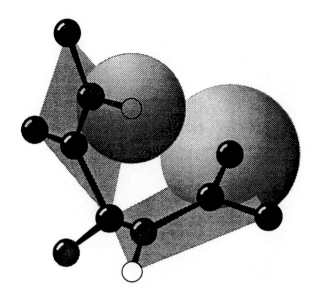

Figure 6-5. Steric interference between adjacent peptide groups.

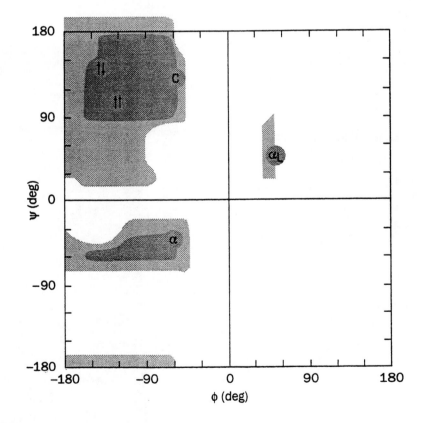

Figure 6-6. A Ramachandran diagram.

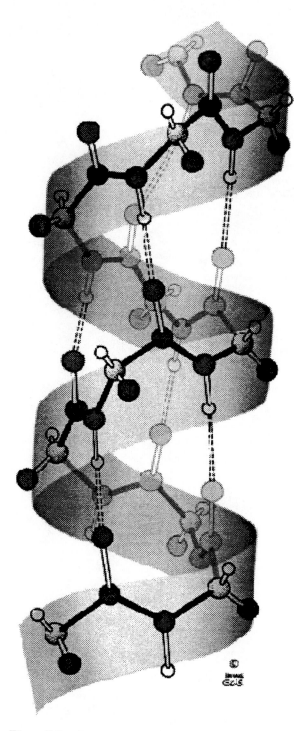

Figure 6-7. **Key to Structure.** The α helix.

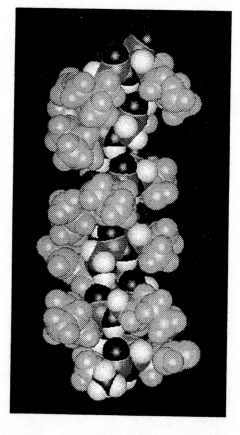

Figure 6-8. **Space-filling model of an α helix.**

(a) **Antiparallel**

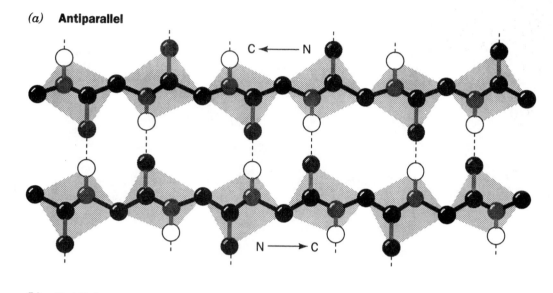

(b) **Parallel**

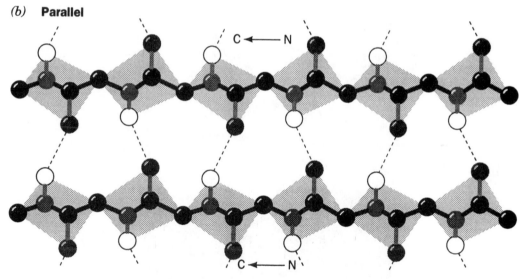

Figure 6-9. *Key to Structure.* β Sheets.

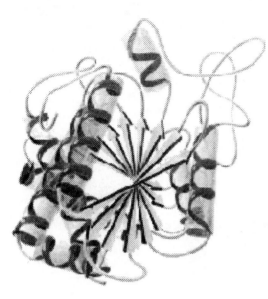

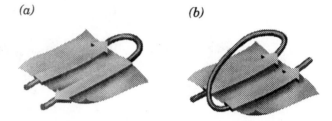

Figure 6-12. Diagram of a β sheet in bovine carboxypeptidase A.

Figure 6-13. Connections between adjacent strands in β sheets.

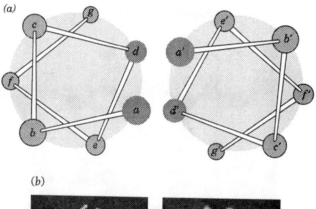

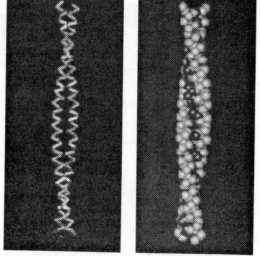

Figure 6-14. The coiled coil of α keratin.

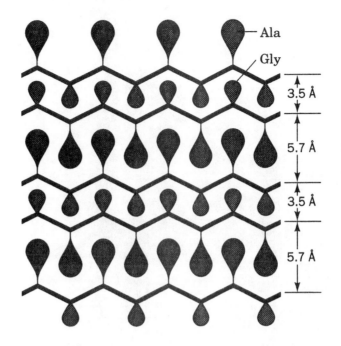

Figure 6-16.   Schematic side view of silk fibroin β sheets.

Figure 6-17.   The collagen triple helix.

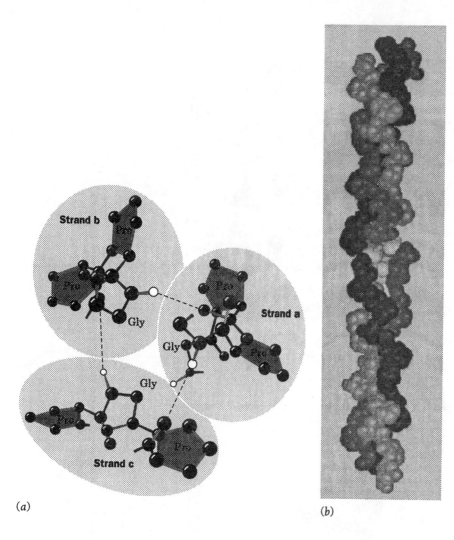

(a)

(b)

Figure 6-18. Molecular interactions in collagen.

Table 6-1. Propensities of Amino Acid Residues for α Helical and β Sheet Conformations

Residue	P_α	P_β
Ala	1.42	0.83
Arg	0.98	0.93
Asn	0.67	0.89
Asp	1.01	0.54
Cys	0.70	1.19
Gln	1.11	1.10
Glu	1.51	0.37
Gly	0.57	0.75
His	1.00	0.87
Ile	1.08	1.60
Leu	1.21	1.30
Lys	1.16	0.74
Met	1.45	1.05
Phe	1.13	1.38
Pro	0.57	0.55
Ser	0.77	0.75
Thr	0.83	1.19
Trp	1.08	1.37
Tyr	0.69	1.47
Val	1.06	1.70

Source: Chou, P.Y. and Fasman, G.D., *Annu. Rev. Biochem.* **47**, 258 (1978).

(a) **Type I**   (b) **Type II**

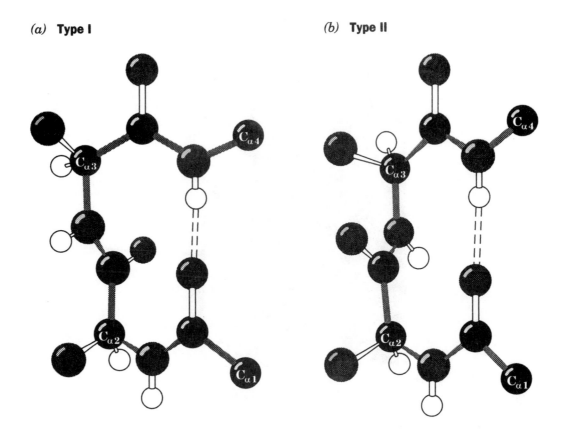

Figure 6-20. Reverse turns in polypeptide chains.

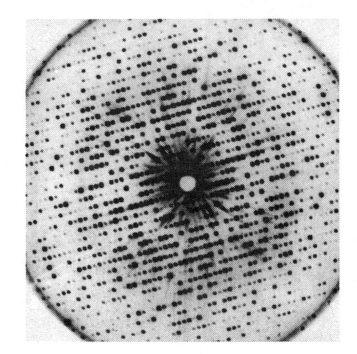

Figure 6-22. An X-ray diffraction photograph of a crystal of sperm whale myoglobin.

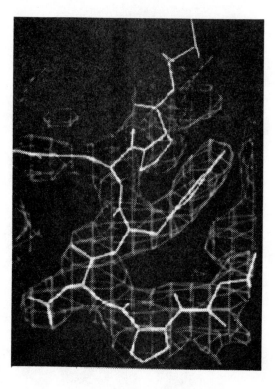

Figure 6-23. An electron density map.

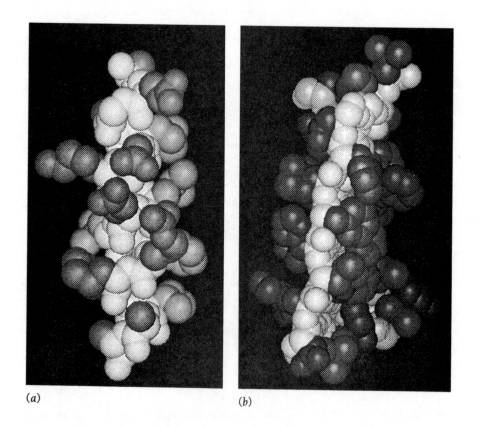

(a)

(b)

Figure 6-25. Side chain locations in an α helix and a β sheet.

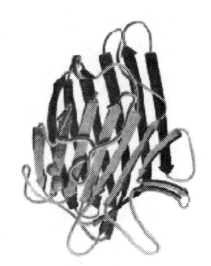
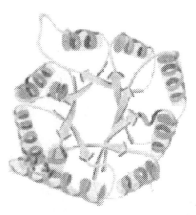

Figure 6-27. Examples of globular proteins.

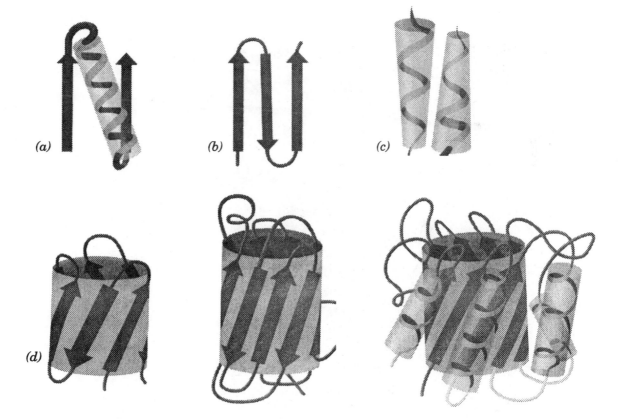

(a)

(b)

(c)

(d)

Figure 6-28. Protein motifs.

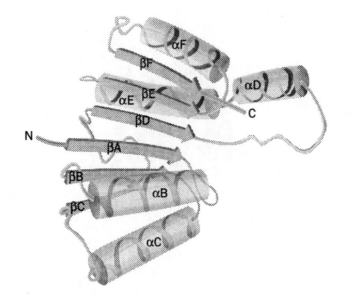

Figure 6-29. An idealized dinucleotide-binding (Rossmann) fold.

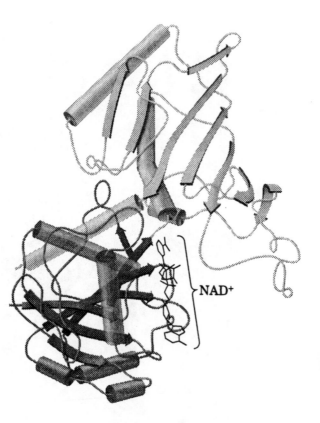

Figure 6-30. The two-domain protein glyceraldehyde-3-phosphate dehydrogenase.

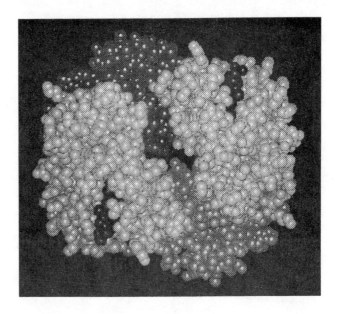

Figure 6-32. Quaternary structure of hemoglobin.

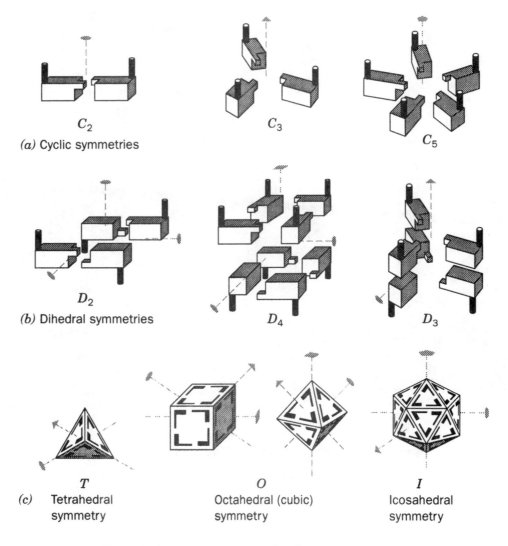

C_2

(a) Cyclic symmetries

C_3

C_5

D_2

(b) Dihedral symmetries

D_4

D_3

T

(c) Tetrahedral symmetry

O

Octahedral (cubic) symmetry

I

Icosahedral symmetry

Figure 6-33. Some symmetries for oligomeric proteins.

Table 6-2. **Hydropathy Scale for Amino Acid Side Chains**

Side Chain	Hydropathy
Ile	4.5
Val	4.2
Leu	3.8
Phe	2.8
Cys	2.5
Met	1.9
Ala	1.8
Gly	−0.4
Thr	−0.7
Ser	−0.8
Trp	−0.9
Tyr	−1.3
Pro	−1.6
His	−3.2
Glu	−3.5
Gln	−3.5
Asp	−3.5
Asn	−3.5
Lys	−3.9
Arg	−4.5

Source: Kyte, J. and Doolittle, R.F., *J. Mol. Biol.* 157, 110 (1982).

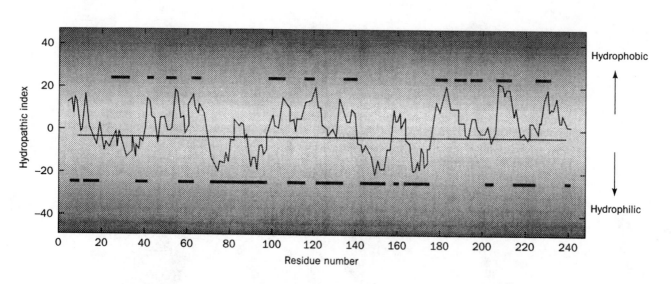

Figure 6-34. **A hydropathic index plot for bovine chymotrypsinogen.**

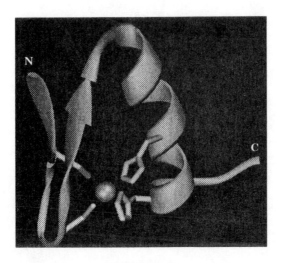

Figure 6-35. A zinc finger motif.

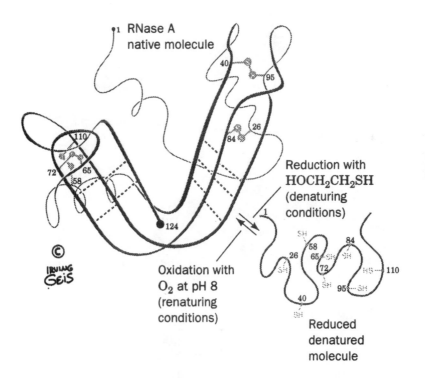

Figure 6-36. The reductive denaturation and oxidative renaturation of RNase A.

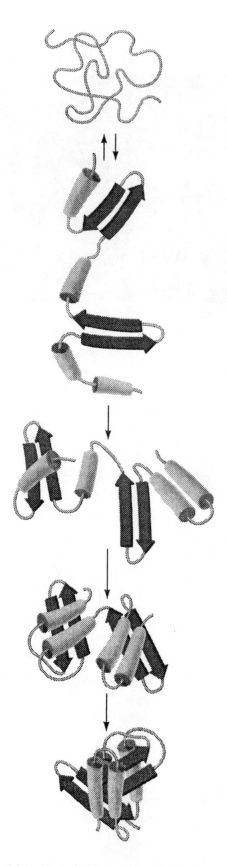

Figure 6-37. **Hypothetical protein folding pathway.**

(a)

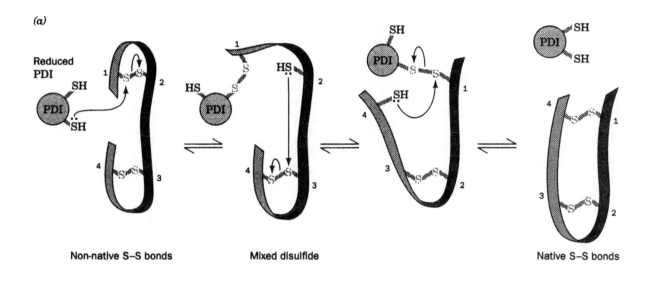

Reduced PDI

Non-native S–S bonds   Mixed disulfide   Native S–S bonds

(b)

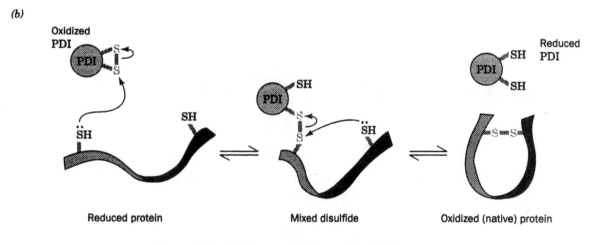

Reduced protein   Mixed disulfide   Oxidized (native) protein

Figure 6-39. **Mechanism of protein disulfide isomerase.**

CHAPTER 7

PROTEIN FUNCTION

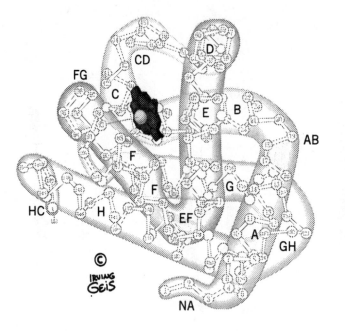

Figure 7-1. **Structure of sperm whale myoglobin.**

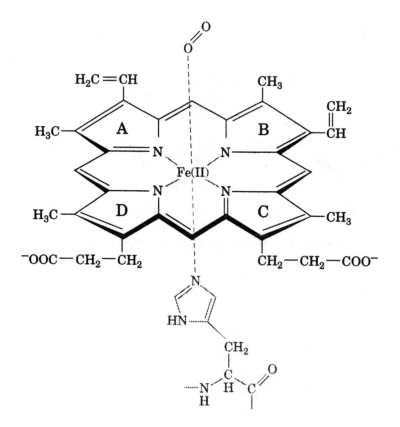

Figure 7-2. **The heme group.**

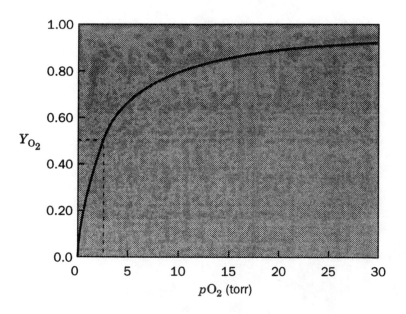

Figure 7-4. Oxygen binding curve of myoglobin.

$$Mb + O_2 \rightleftharpoons MbO_2$$

$$K = \frac{[Mb][O_2]}{[MbO_2]} \qquad [7\text{-}1]$$

$$Y_{O_2} = \frac{[MbO_2]}{[Mb] + [MbO_2]} = \frac{[O_2]}{K + [O_2]} \qquad [7\text{-}2]$$

$$Y_{O_2} = \frac{pO_2}{K + pO_2} \qquad [7\text{-}3]$$

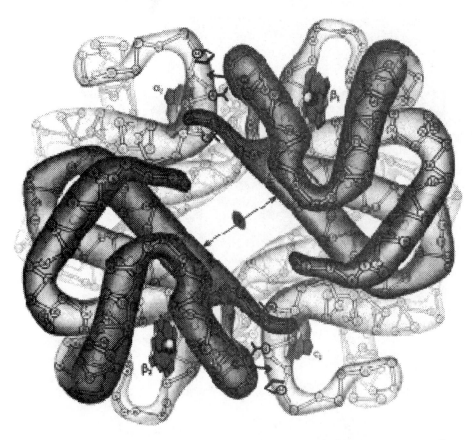

(a)

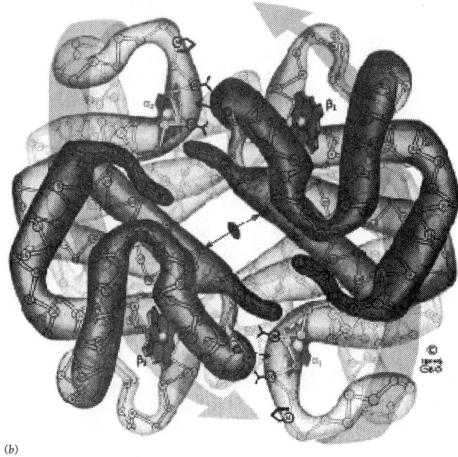

(b)

Figure 7-5. Hemoglobin structure.

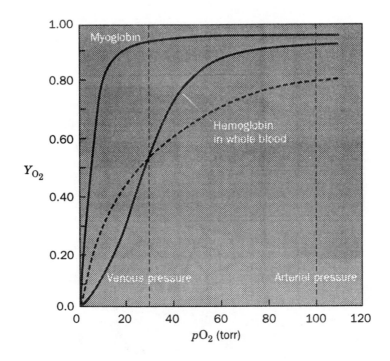

Figure 7-7. *Key to Function.* **Oxygen binding curve of hemoglobin.**

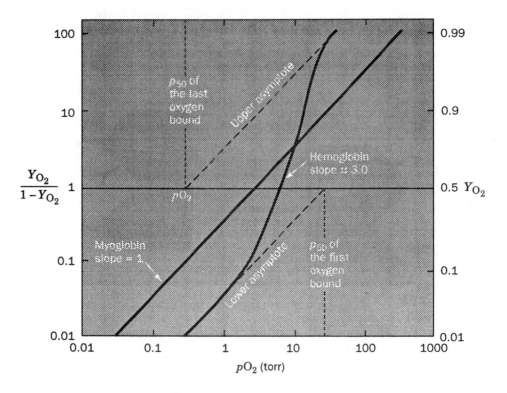

Figure 7-8. **Hill plots for myoglobin and purified hemoglobin.**

$$\log\left(\frac{Y_{O_2}}{1 - Y_{O_2}}\right) = n \log pO_2 - n \log p_{50} \qquad [7\text{-}6]$$

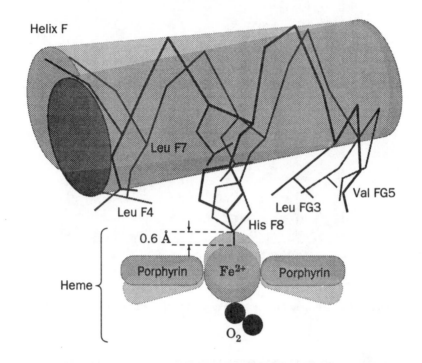

Figure 7-9. Movements of the heme and the F helix during the T → R transition in hemoglobin.

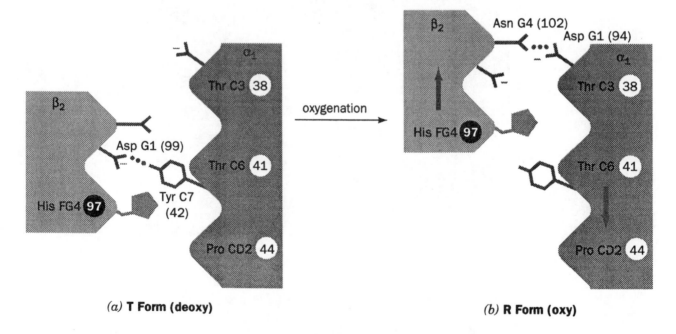

(a) **T Form (deoxy)**　　　　　　　　(b) **R Form (oxy)**

Figure 7-10. Changes at the α_1–β_2 interface during the T → R transition in hemoglobin.

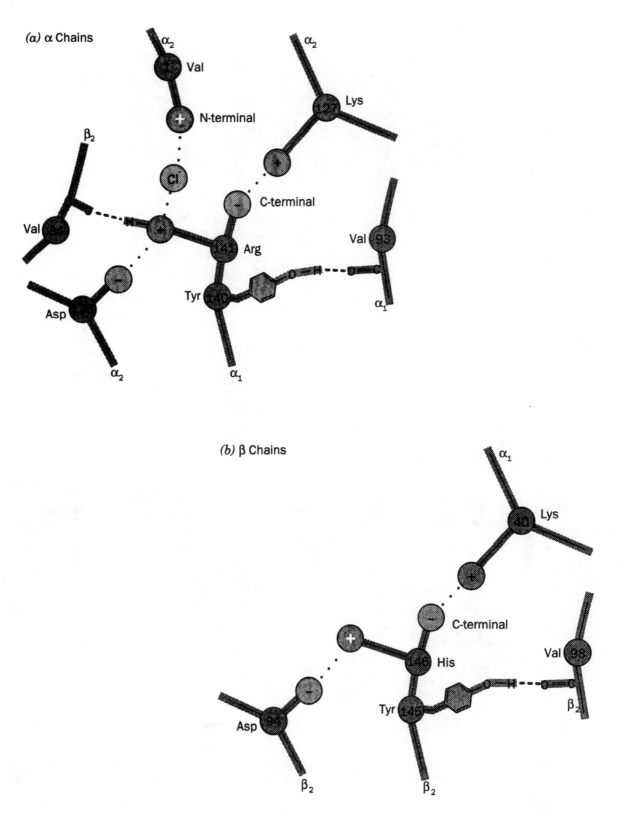

(a) α Chains

(b) β Chains

Figure 7-11. Networks of ion pairs and hydrogen bonds in deoxyhemoglobin.

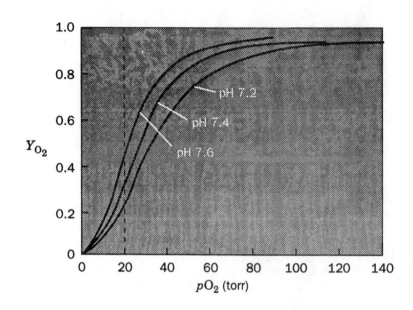

Figure 7-12. The Bohr effect.

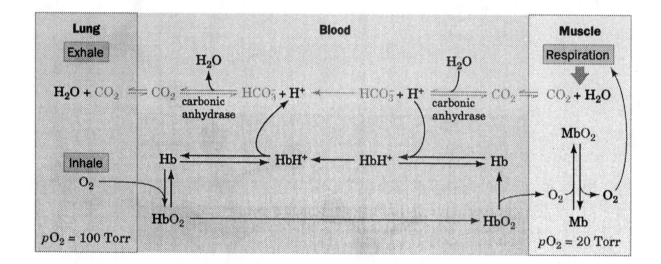

Figure 7-13. *Key to Function*. The roles of hemoglobin and myoglobin in transporting O_2 from the lungs to respiring tissues and CO_2 (as HCO_3^-) from the tissues to the lungs.

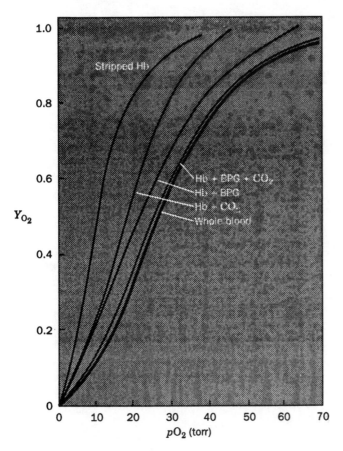

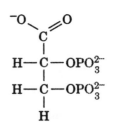

D-2,3-Bisphosphoglycerate (BPG)

Figure 7-14. The effects of BPG and CO₂ on hemoglobin's O₂ dissociation curve.

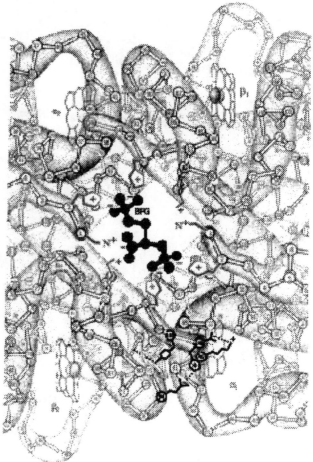

Figure 7-15. Binding of BPG to deoxyhemoglobin.

Table 7-1. **Some Hemoglobin Variants**

Name[a]	Mutation	Effect
Hammersmith	Phe CD1(42)β → Ser	Weakens heme binding
Bristol	Val E11(67)β → Asp	Weakens heme binding
Bibba	Leu H19(136)α → Pro	Disrupts the H helix
Savannah	Gly B6(24)β → Val	Disrupts the B–E helix interface
Philly	Tyr C1(35)α → Phe	Disrupts hydrogen bonding at the α_1–β_1 interface
Boston	His E7(58)α → Tyr	Promotes methemoglobin formation
Milwaukee	Val E11(67)β → Glu	Promotes methemoglobin formation
Iwate	His F8(87)α → Tyr	Promotes methemoglobin formation
Yakima	Asp G1(99)β → His	Disrupts a hydrogen bond that stabilizes the T conformation
Kansas	Asn G4(102)β → Thr	Disrupts a hydrogen bond that stabilizes the R conformation

[a]Hemoglobin variants are usually named after the place where they were discovered (e.g., hemoglobin Hammersmith).

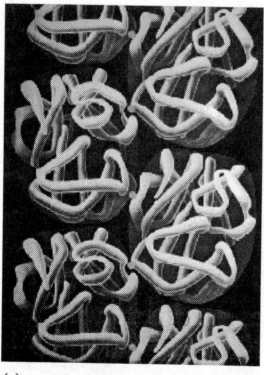

(a)

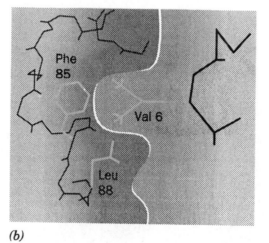

(b)

Figure 7-17. **Structure of a deoxyhemoglobin S fiber.**

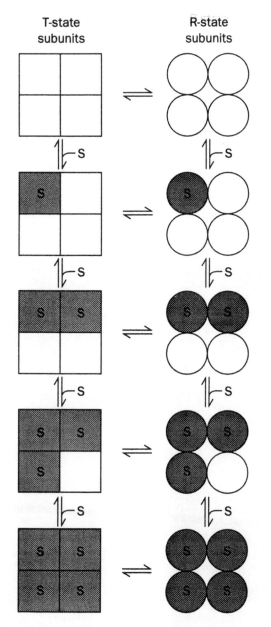

T-state subunits   **R-state subunits**

Figure 7-19. **The symmetry model of allosterism.**

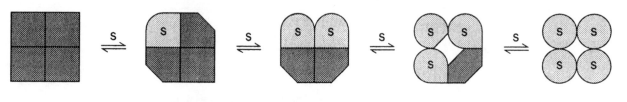

Figure 7-20. **The sequential model of allosterism.**

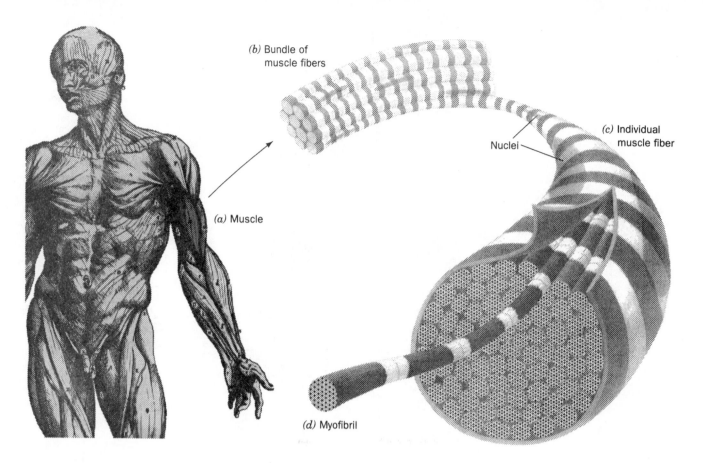

Figure 7-22.   Skeletal muscle organization.

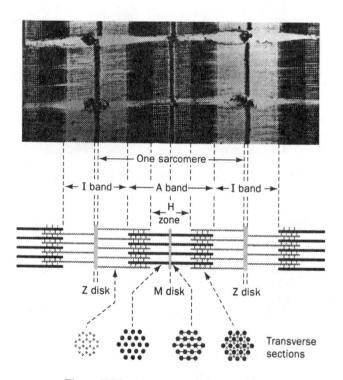

Figure 7-23.   Anatomy of the myofibril.

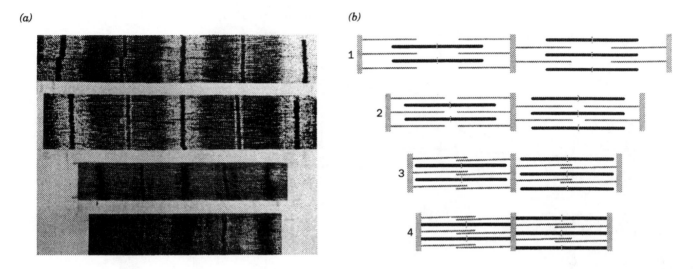

(a)   (b)

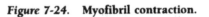

Figure 7-24. Myofibril contraction.

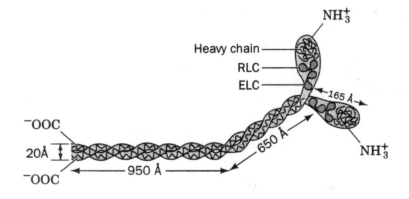

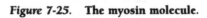

Figure 7-25. The myosin molecule.

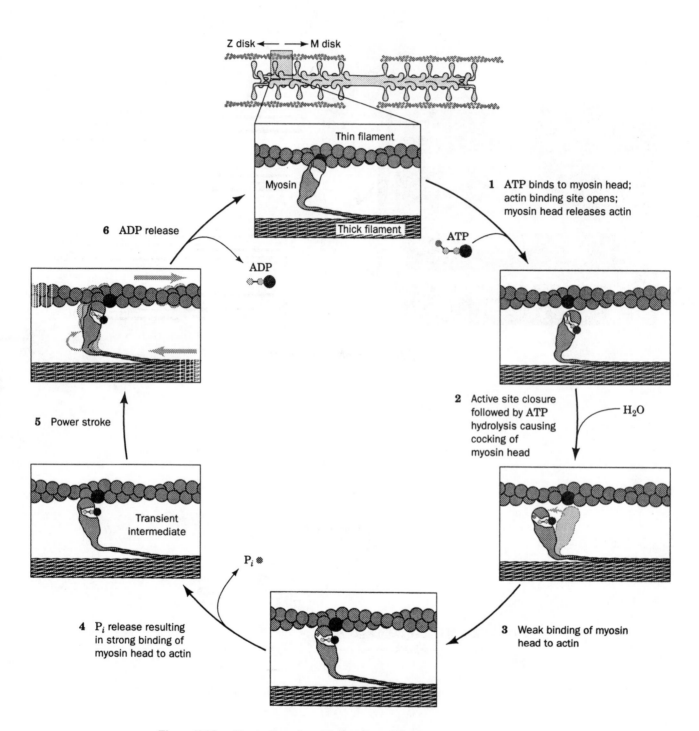

Figure 7-29. *Key to Function.* **Mechanism of force generation in muscle.**

74

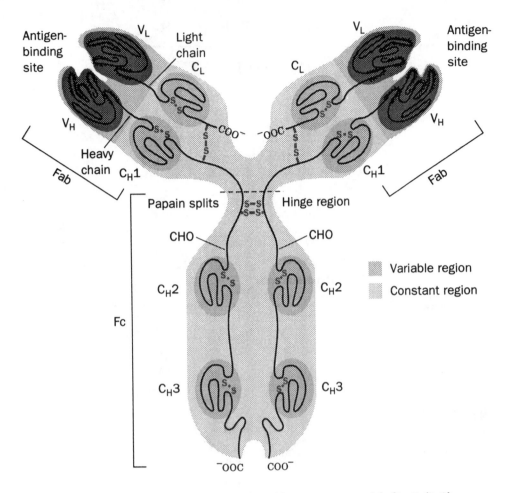

Figure 7-34. *Key to Structure.* Diagram of human immunoglobulin G (IgG).

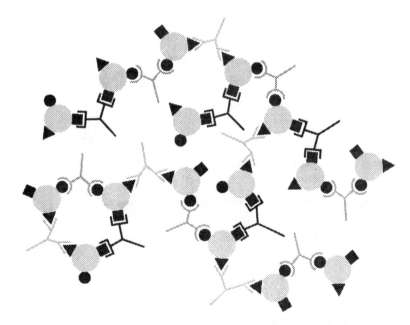

Figure 7-37. Antigen cross-linking by antibodies.

CARBOHYDRATES

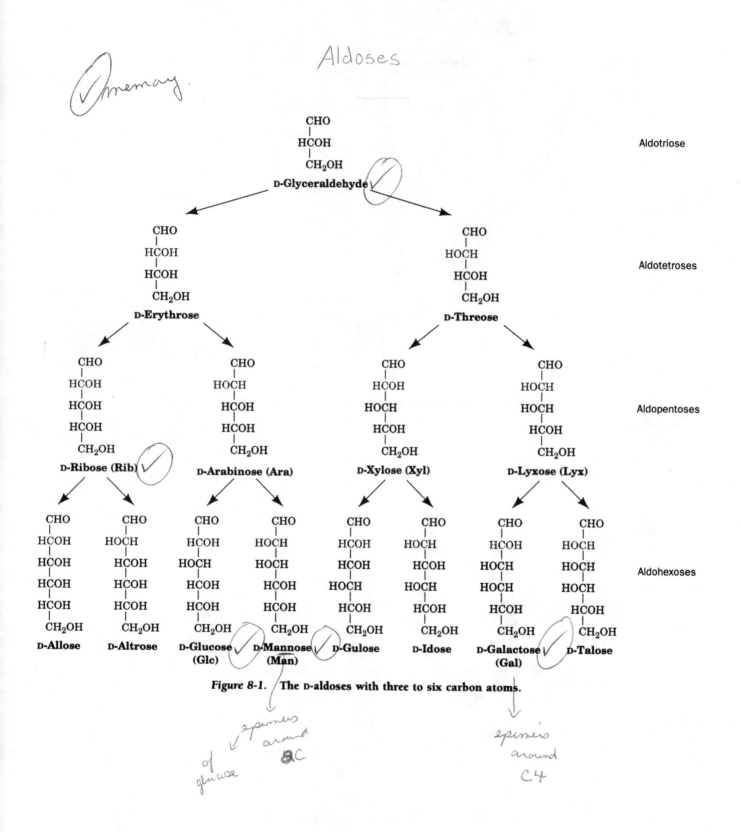

Figure 8-1. The D-aldoses with three to six carbon atoms.

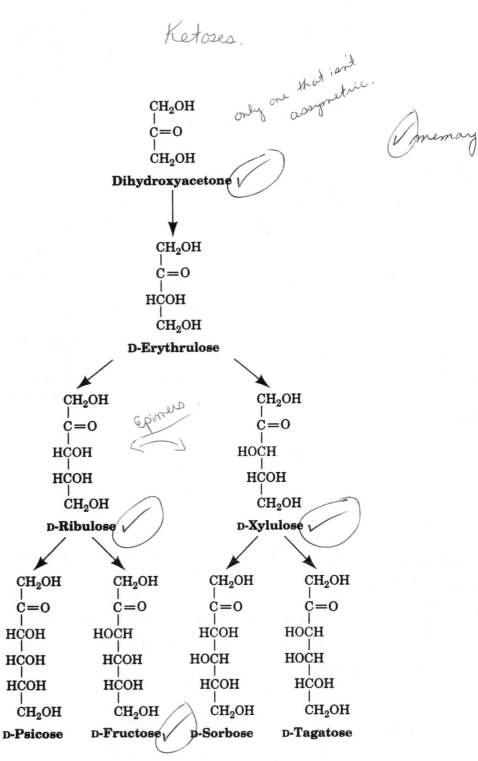

Ketoses.

only one that isn't asymmetric.

✓ memay.

Figure 8-2. The D-ketoses with three to six carbon atoms.

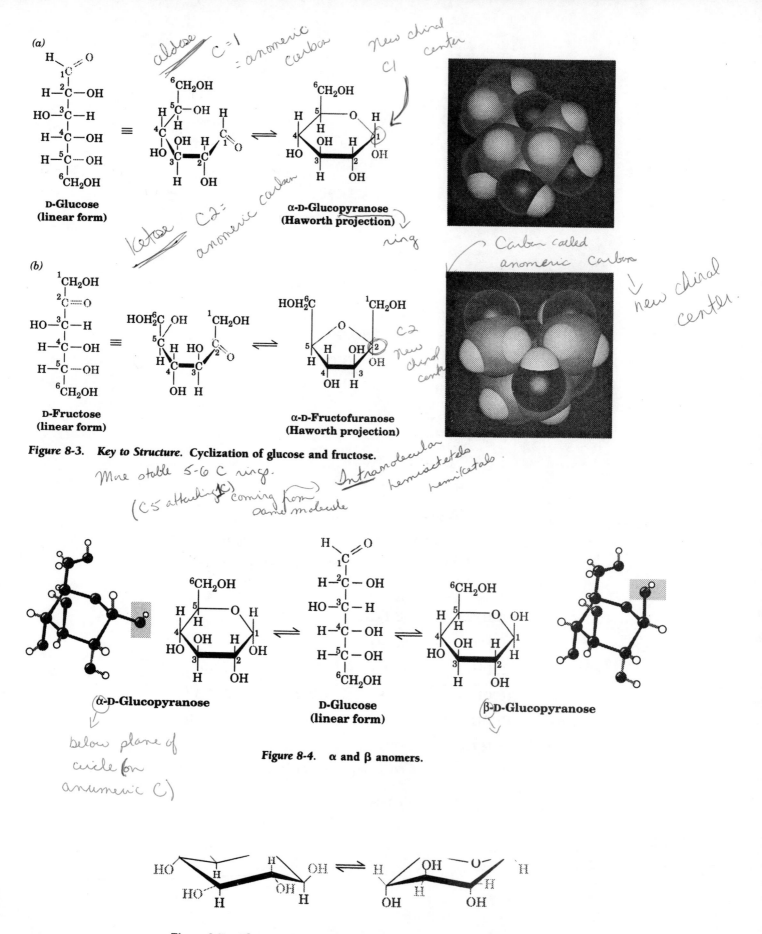

(a)

**D-Glucose
(linear form)**

aldose C=1 = *anomeric carbon*

new chiral center C1

**α-D-Glucopyranose
(Haworth projection)**

ring

Ketose C2 = *anomeric carbon*

Carbon called anomeric carbon

new chiral center.

(b)

**D-Fructose
(linear form)**

**α-D-Fructofuranose
(Haworth projection)**

C2 *new chiral center*

Figure 8-3. Key to Structure. Cyclization of glucose and fructose.

More stable 5-6 C ring.
(C5 attacking 1C) coming from same molecule *Intramolecular hemiacetals hemiketals.*

α-D-Glucopyranose

below plane of circle (on anomeric C)

**D-Glucose
(linear form)**

β-D-Glucopyranose

Figure 8-4. α and β anomers.

Figure 8-5. The two chair conformations of β-D-glucopyranose.

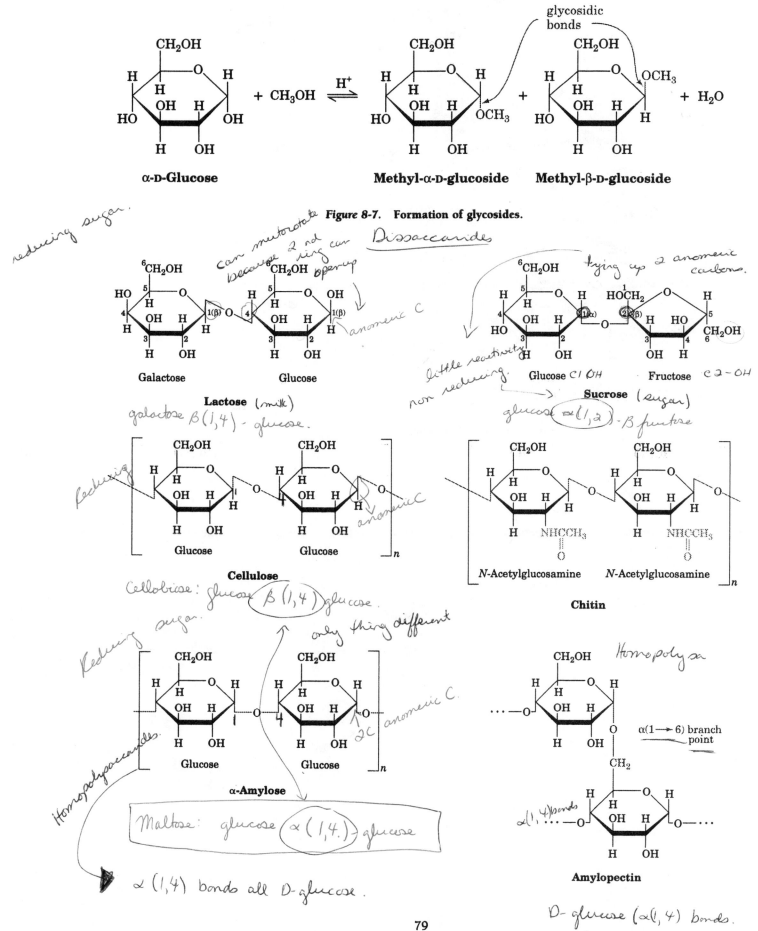

Figure 8-7. Formation of glycosides.

α-D-Glucose + CH₃OH ⇌ Methyl-α-D-glucoside + Methyl-β-D-glucoside + H₂O

glycosidic bonds

reducing sugar.

can mutarotate because 2nd ring can open up

Dissaccarides

anomeric C

Galactose   Glucose

Lactose (milk)

galactose β(1,4)-glucose.

tying up 2 anomeric carbons.

little reactivity non reducing.

Glucose C1 OH   Fructose C2-OH

Sucrose (sugar)

glucose α(1,2)-β fructose

Reducing

Glucose   Glucose   anomeric C

Cellulose

Cellobiose: glucose β(1,4) glucose.

Reducing sugar.

N-Acetylglucosamine   N-Acetylglucosamine

Chitin

only thing different

Homopolysaccarides.

2C anomeric C.

Glucose   Glucose

α-Amylose

Homopoly sa

α(1 → 6) branch point

α(1,4) bonds

Amylopectin

Maltose: glucose α(1,4) glucose

α(1,4) bonds all D-glucose.

D-glucose (α(1,4) bonds. Branching α(1,6) bonds.

79

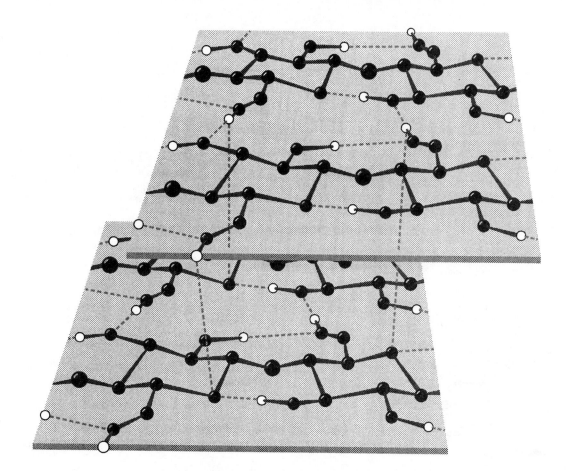

Figure 8-9. Model of cellulose.

α (1,4) bond has slight twist to it. linear chain forms helical structure.

Figure 8-10. α-Amylose.

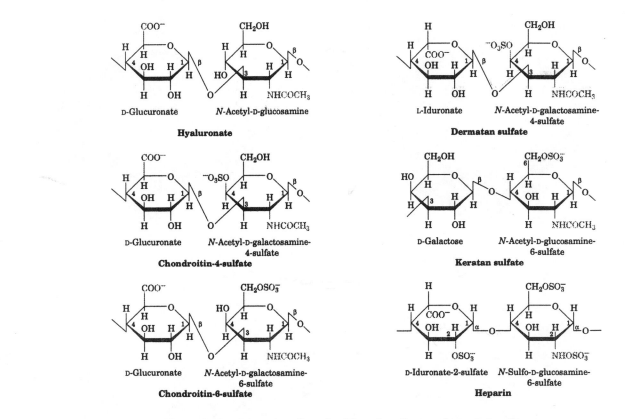

Figure 8-12. Repeating disaccharide units of some glycosaminoglycans.

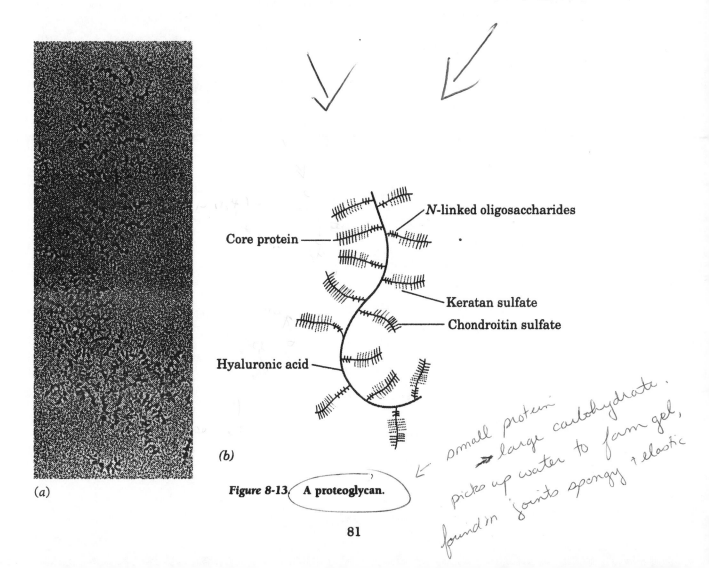

Core protein

N-linked oligosaccharides

Keratan sulfate

Chondroitin sulfate

Hyaluronic acid

(b)

(a)

Figure 8-13. A proteoglycan.

← small protein
→ large carbohydrate.
picks up water to form gel,
found in joints spongy + elastic

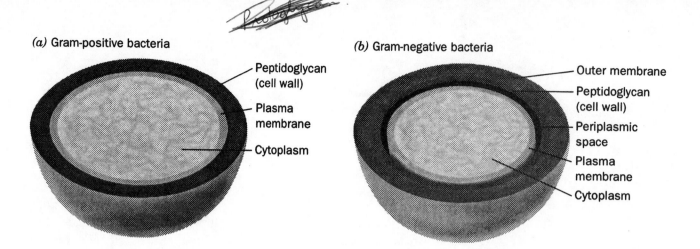

(a) Gram-positive bacteria

- Peptidoglycan (cell wall)
- Plasma membrane
- Cytoplasm

(b) Gram-negative bacteria

- Outer membrane
- Peptidoglycan (cell wall)
- Periplasmic space
- Plasma membrane
- Cytoplasm

Figure 8-14. Bacterial cell walls.

Heterosaccarides.

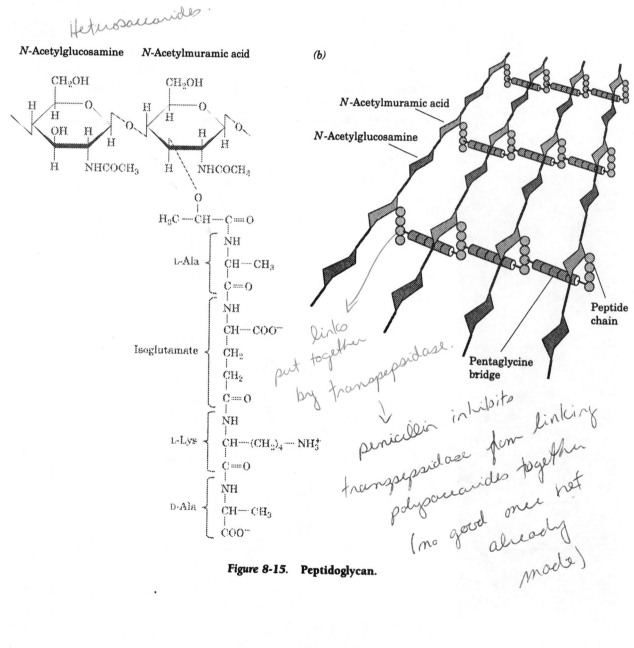

(a)

N-Acetylglucosamine   N-Acetylmuramic acid

L-Ala
Isoglutamate
L-Lys
D-Ala

(b)

N-Acetylmuramic acid
N-Acetylglucosamine

Peptide chain
Pentaglycine bridge

links put together by transpepsidase.

penicillin inhibits transpepsidase from linking polysaccarides together (no good once net already made)

Figure 8-15. Peptidoglycan.

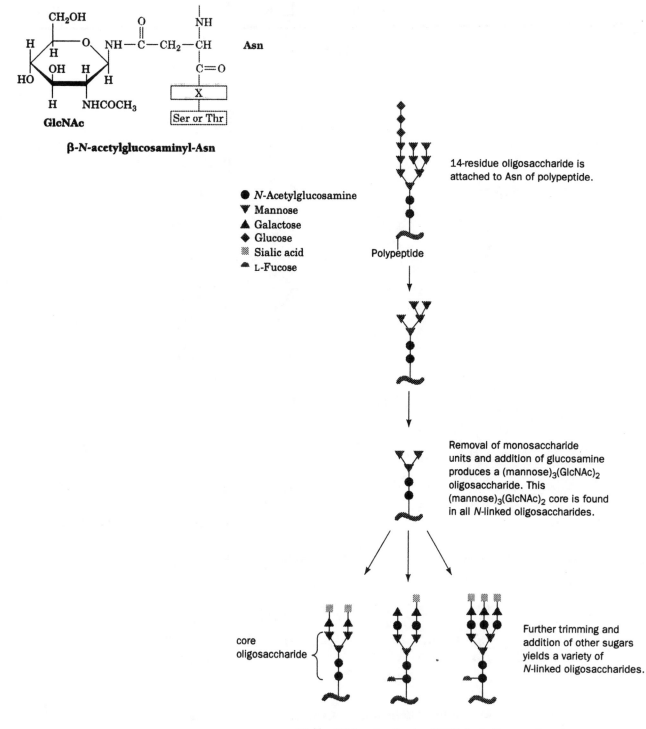

GlcNAc

β-N-acetylglucosaminyl-Asn

Asn

● *N*-Acetylglucosamine
▼ Mannose
▲ Galactose
◆ Glucose
▨ Sialic acid
▬ L-Fucose

Polypeptide

14-residue oligosaccharide is attached to Asn of polypeptide.

Removal of monosaccharide units and addition of glucosamine produces a (mannose)$_3$(GlcNAc)$_2$ oligosaccharide. This (mannose)$_3$(GlcNAc)$_2$ core is found in all *N*-linked oligosaccharides.

core oligosaccharide

Further trimming and addition of other sugars yields a variety of *N*-linked oligosaccharides.

Figure 8-16. **Synthesis of N-linked oligosaccharides.**

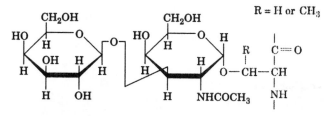

R = H or CH$_3$

β-Galactosyl-(1 ⟶ 3)-α-*N*-acetylgalactosyl-Ser/Thr

CHAPTER 9

LIPIDS

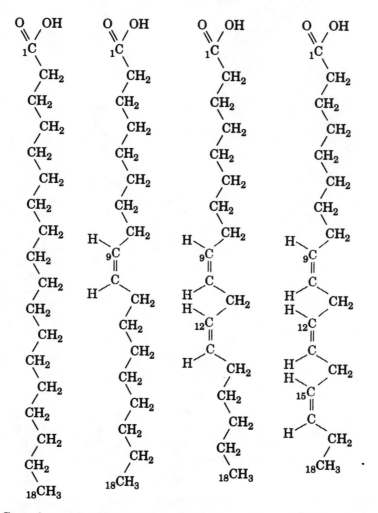

Stearic acid   Oleic acid   Linoleic acid   α-Linolenic acid

Figure 9-1. The structural formulas of some C_{18} fatty acids.

Table 9-2. The Common Classes of Glycerophospholipids

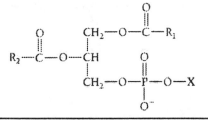

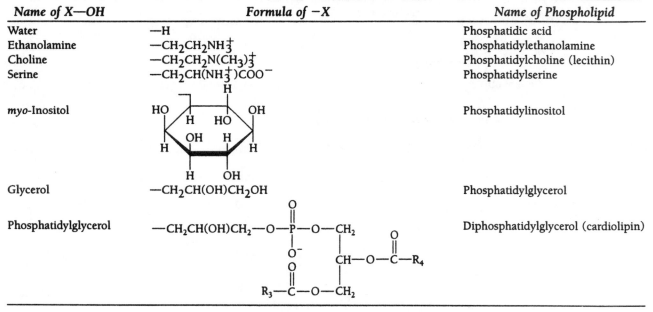

Name of X—OH	Formula of −X	Name of Phospholipid
Water	—H	Phosphatidic acid
Ethanolamine	$-CH_2CH_2NH_3^+$	Phosphatidylethanolamine
Choline	$-CH_2CH_2N(CH_3)_3^+$	Phosphatidylcholine (lecithin)
Serine	$-CH_2CH(NH_3^+)COO^-$	Phosphatidylserine
myo-Inositol		Phosphatidylinositol
Glycerol	$-CH_2CH(OH)CH_2OH$	Phosphatidylglycerol
Phosphatidylglycerol		Diphosphatidylglycerol (cardiolipin)

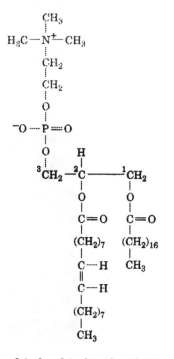

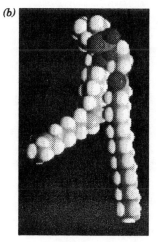

1-Stearoyl-2-oleoyl-3-phosphatidylcholine

Figure 9-4. The glycerophospholipid 1-stearoyl-2-oleoyl-3-phosphatidylcholine.

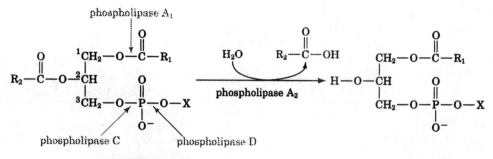

Figure 9-5. Action of phospholipases.

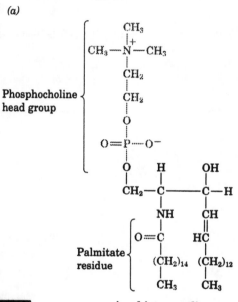

A sphingomyelin

Figure 9-7. A sphingomyelin.

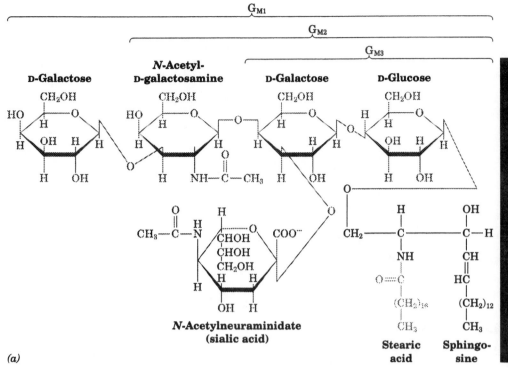

G_{M1}

G_{M2}

G_{M3}

D-Galactose

N-Acetyl-D-galactosamine

D-Galactose

D-Glucose

CH₂OH

CH₂OH

CH₂OH

CH₂OH

HO

HO

H

H

OH

H

OH

H

H

OH

H

H

OH

H

OH

NH—C—CH₃

H

OH

CH₃—C—N

CHOH

CHOH

CH₂OH

H

H

OH

H

COO⁻

O

CH₂

C

C—H

NH

OH

O=C

CH

(CH₂)₁₆

HC

(CH₂)₁₂

CH₃

CH₃

**N-Acetylneuraminidate
(sialic acid)**

**Stearic
acid**

**Sphingo-
sine**

(a)

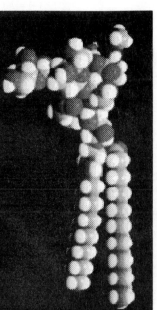

(b)

Figure 9-9. Gangliosides.

(a)

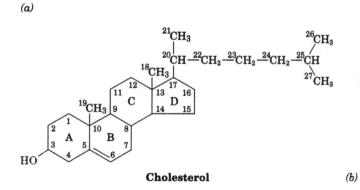

²¹CH₃

²⁰CH—²²CH₂—²³CH₂—²⁴CH₂—²⁵CH

¹⁸CH₃

²⁶CH₃

²⁷CH₃

12

11

C

17

13

16

19CH₃

9

D

1

10

8

14

15

2

A

B

7

3

5

6

HO

4

Cholesterol

(b)

Figure 9-10. Cholesterol.

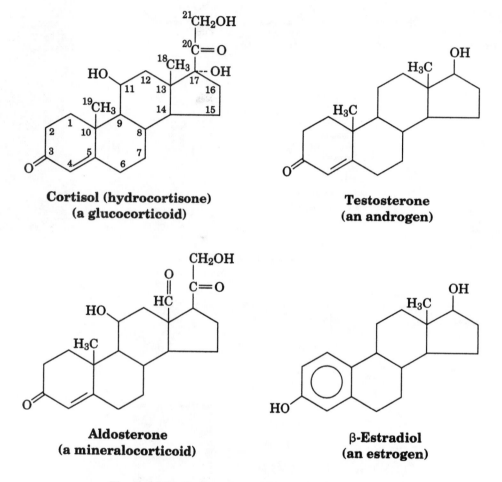

Cortisol (hydrocortisone)
(a glucocorticoid)

Testosterone
(an androgen)

Aldosterone
(a mineralocorticoid)

β-Estradiol
(an estrogen)

Figure 9-11. **Some representative steroid hormones.**

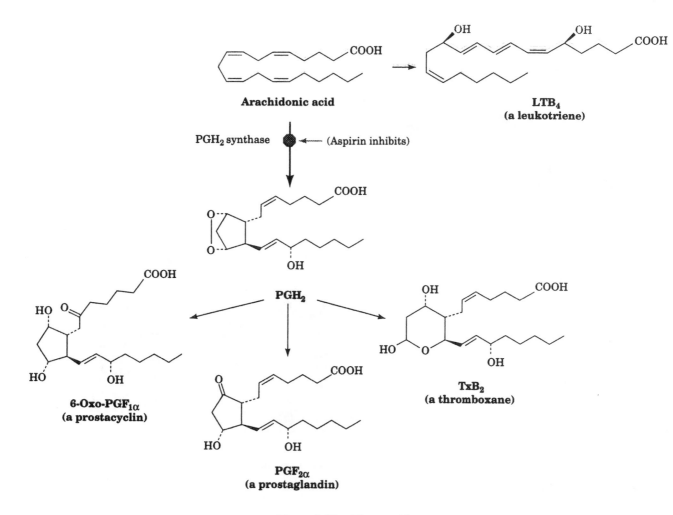

Arachidonic acid

LTB₄
(a leukotriene)

PGH₂ synthase   ←——— (Aspirin inhibits)

PGH₂

6-Oxo-PGF₁ₐ
(a prostacyclin)

PGF₂ₐ
(a prostaglandin)

TxB₂
(a thromboxane)

Figure 9-12.   Eicosanoids.

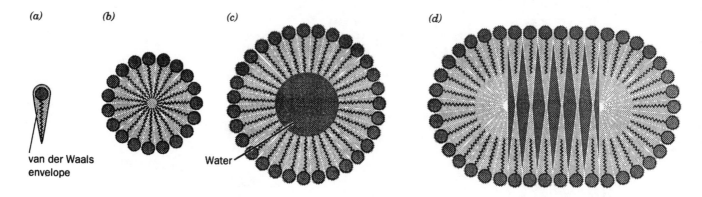

(a)   *(b)*   *(c)*   *(d)*

van der Waals
envelope

Water

Figure 9-13.   Aggregates of single-tailed lipids.

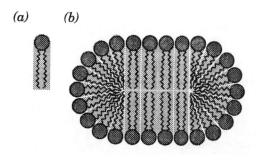

(a) *(b)*

Figure 9-14. **Bilayer formation by phospholipids.**

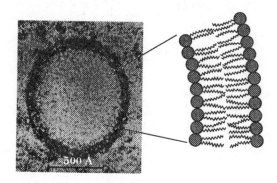

Figure 9-15. **Electron micrograph of a liposome.**

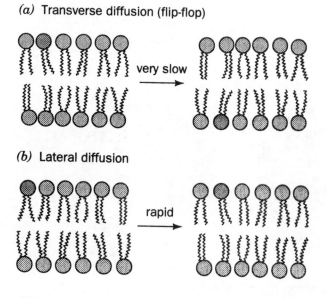

(a) Transverse diffusion (flip-flop)

very slow

(b) Lateral diffusion

rapid

Figure 9-16. **Phospholipid diffusion in a lipid bilayer.**

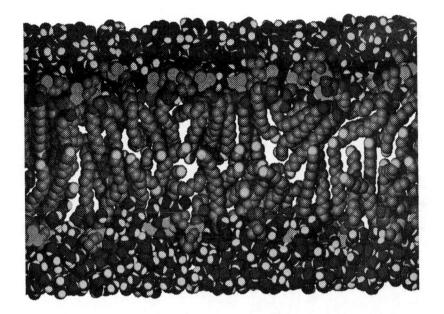

Figure 9-17. Model (snapshot) of a lipid bilayer at a particular instant in time.

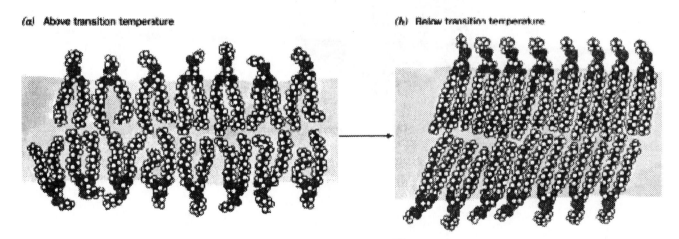

(a) Above transition temperature   *(b) Below transition temperature*

Figure 9-18. Phase transition in a lipid bilayer.

CHAPTER 10

BIOLOGICAL MEMBRANES

Figure 10-1. Model of an integral membrane protein.

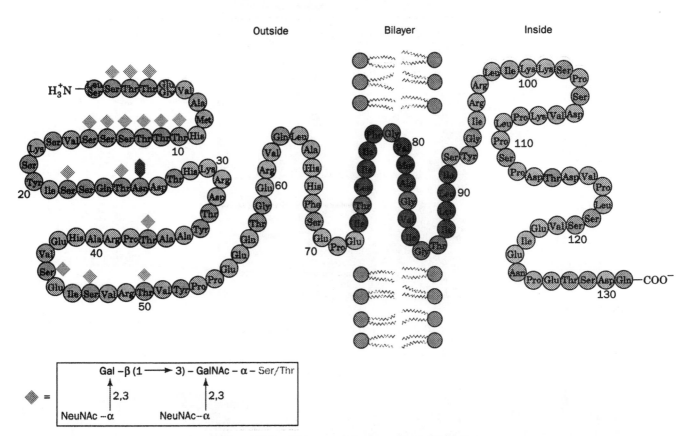

Outside　　　　　　Bilayer　　　　　　Inside

$$Gal - \beta \, (1 \longrightarrow 3) - GalNAc - \alpha - Ser/Thr$$

◈ =

$$
\begin{array}{ccc}
& 2,3 & 2,3 \\
NeuNAc - \alpha & & NeuNAc - \alpha
\end{array}
$$

Figure 10-2. **Human erythrocyte glycophorin A.**

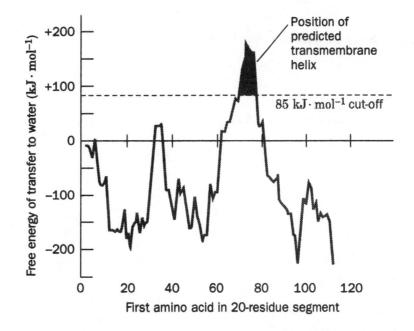

Figure 10-3. **Identification of glycophorin A's transmembrane domain.**

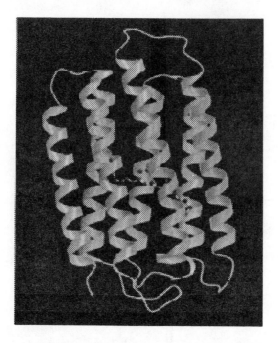

Figure 10-4. The structure of bacteriorhodopsin.

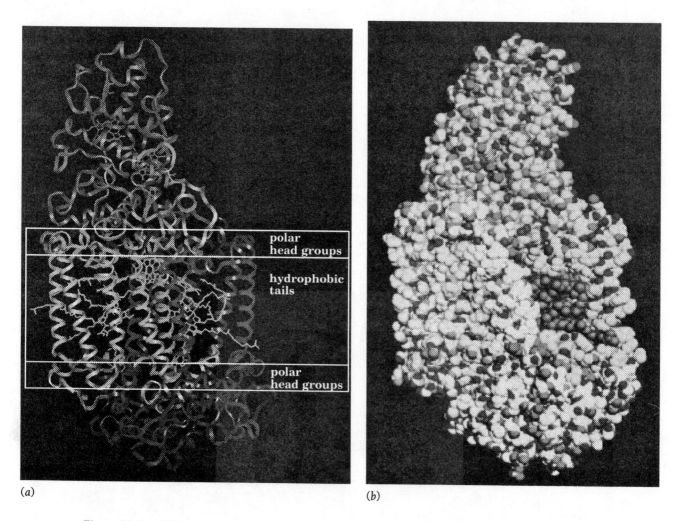

polar
head groups

hydrophobic
tails

polar
head groups

(a)

(b)

Figure 10-5. X-Ray structure of the photosynthetic reaction center of *Rhodopseudomonas viridis*.

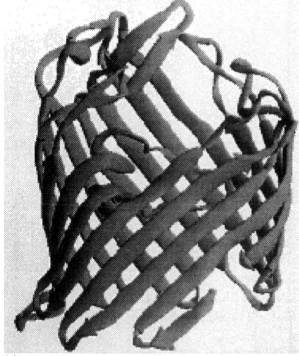

(a)

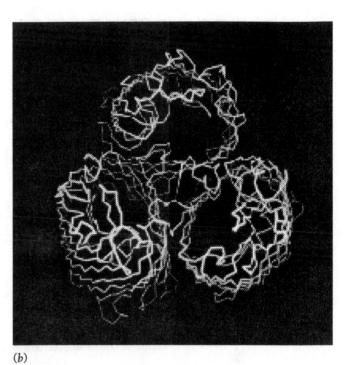

(b)

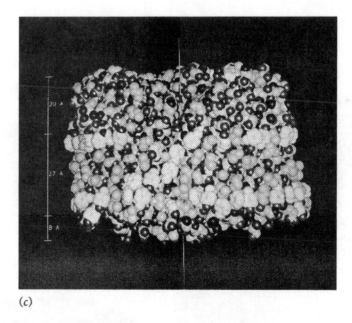

(c)

Figure 10-6. **X-Ray structure of the *E. coli* OmpF porin.**

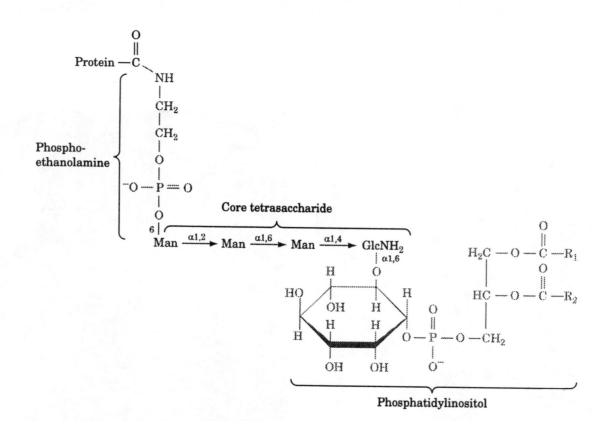

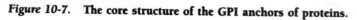

Figure 10-7. **The core structure of the GPI anchors of proteins.**

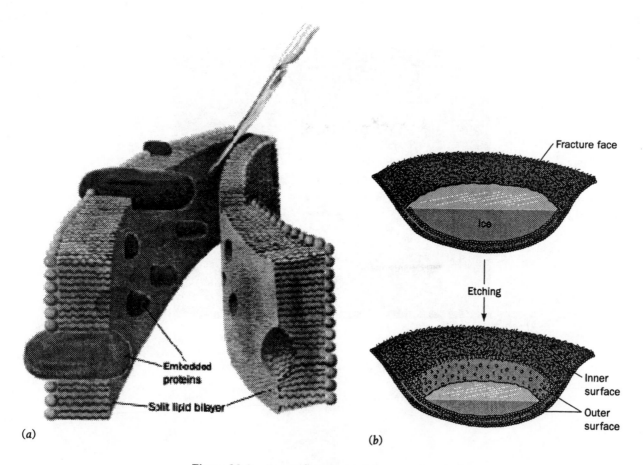

Figure 10-8. **Freeze-fracture and freeze-etch techniques.**

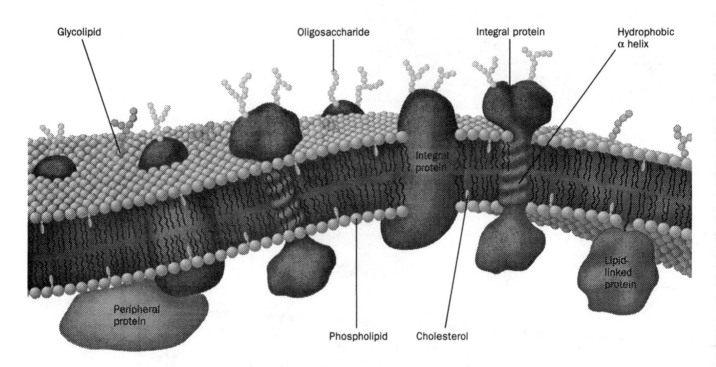

Glycolipid

Oligosaccharide

Integral protein

Hydrophobic
α helix

Integral
protein

Phospholipid

Cholesterol

Lipid-
linked
protein

Peripheral
protein

Figure 10-10. **Diagram of a plasma membrane.**

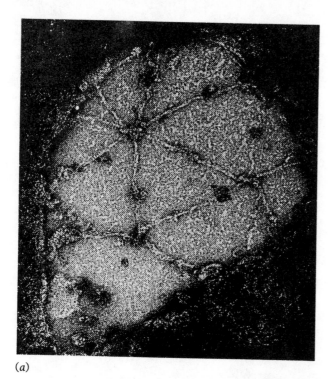

(a)

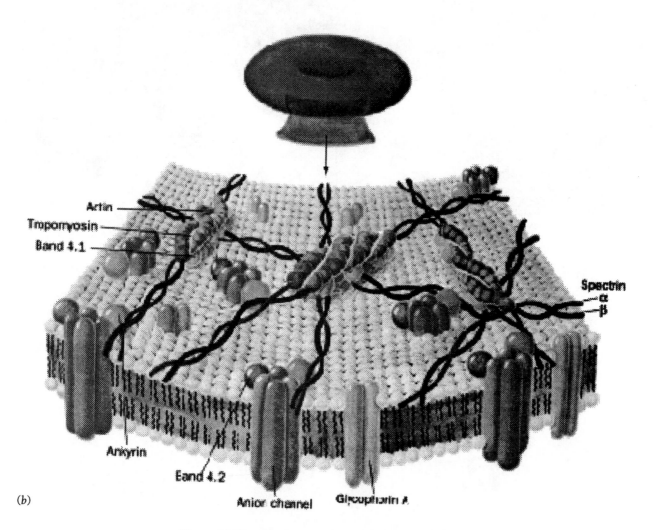

Actin

Tropomyosin

Band 4.1

Spectrin
α
β

Ankyrin

Band 4.2

Anion channel

Glycophorin A

(b)

Figure 10-15. The human erythrocyte membrane skeleton.

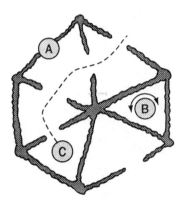

Figure 10-16. Model rationalizing the various mobilities of membrane proteins.

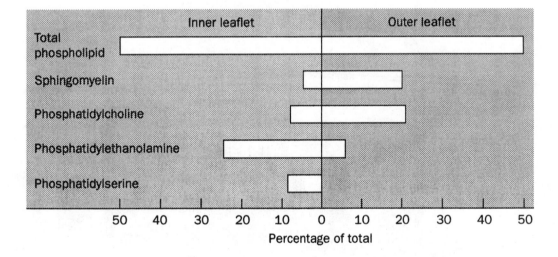

Figure 10-17. Asymmetric distribution of membrane phospholipids in the human erythrocyte membrane.

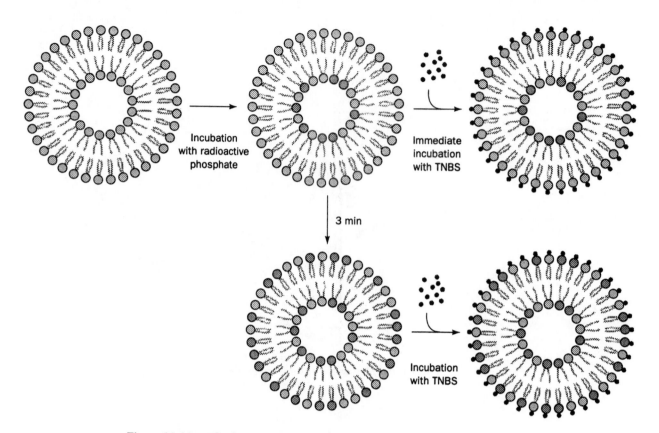

Figure 10-19. The location of lipid synthesis in a bacterial membrane.

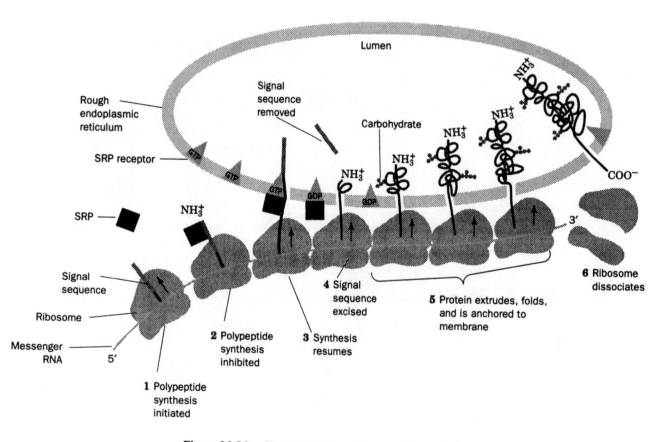

Figure 10-20. *Key to Function.* The signal hypothesis.

Bovine growth hormone	M M A A G P R T S ▓▓▓▓▓▓▓▓ W T Q V V G	A F P
Bovine proalbumin	M K W V T ▓▓▓▓▓ S S A Y S	R G V
Human proinsulin	M A L W M R ▓▓▓▓▓▓▓ W G P D P A A A	F V N
Human interferon-γ	M K Y T S Y ▓▓▓▓▓▓▓ G S L G	C Y C
Human α-fibrinogen	M F S M R ▓▓▓▓▓▓▓ T A W T	A D S
Human IgG heavy chain	M E F G L S W ▓▓▓▓▓ K G V Q C	E V Q
Rat amylase	M K ▓▓▓▓▓▓▓ C W A	Q Y D
Murine α-fetoprotein	M K W I T P A S ▓▓▓▓▓ H F A A S K	A L H
Chicken lysozyme	M R S ▓▓▓▓▓▓▓▓ G	K V F
Zea mays rein protein 22.1	M A T K ▓▓▓▓▓▓▓▓ S A T N A	F I I

Figure 10-21.   The N-terminal sequences of some eukaryotic secretory preproteins.

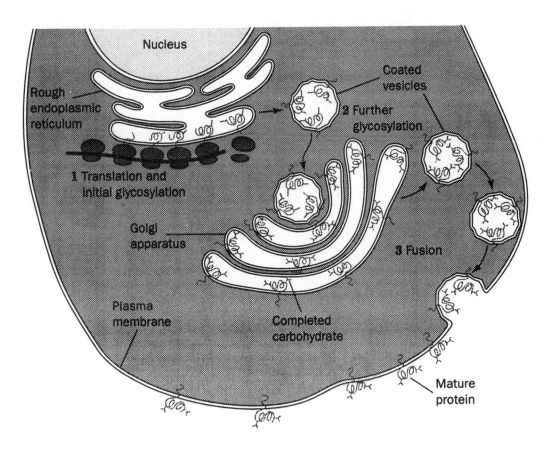

Figure 10-22.   *Key to Function.*   The post-translational processing of integral membrane proteins.

101

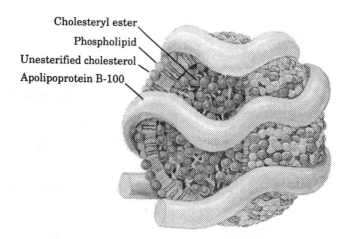

Cholesteryl ester
Phospholipid
Unesterified cholesterol
Apolipoprotein B-100

Figure 10-25. **Diagram of LDL, the major cholesterol carrier of the bloodstream.**

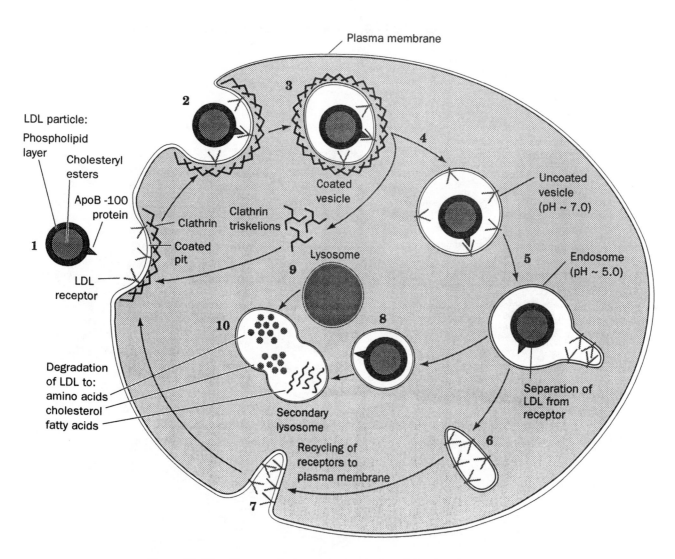

Figure 10-27. *Key to Function.* **Receptor-mediated endocytosis of LDL.**

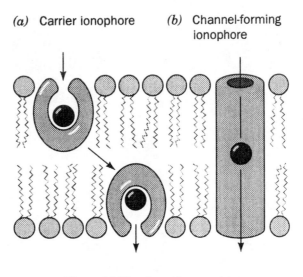

(a) Carrier ionophore   *(b)* Channel-forming ionophore

Figure 10-29.   **Ionophore action.**

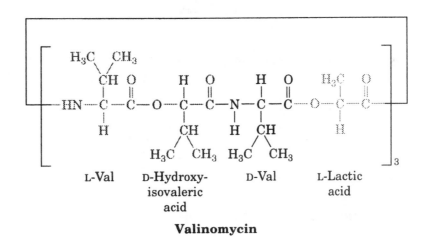

L-Val   D-Hydroxy-   D-Val   L-Lactic
         isovaleric   acid
         acid

Valinomycin

Figure 10-30.   **Valinomycin.**

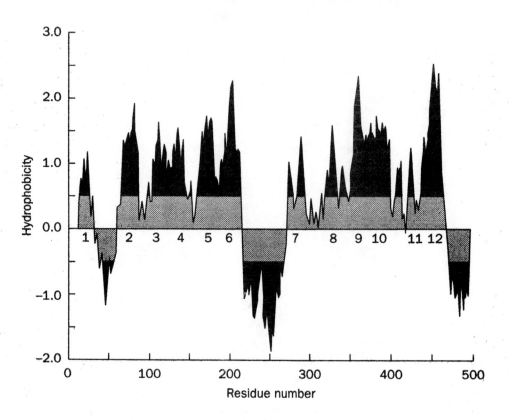

Figure 10-34. Hydropathy plot of the human erythrocyte glucose transporter.

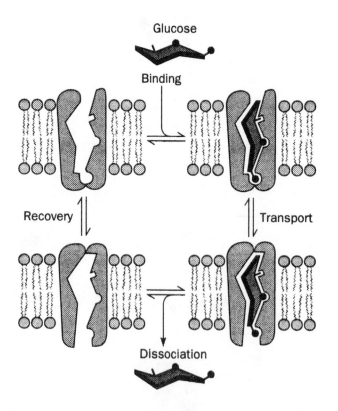

Figure 10-35. *Key to Function.* Model for glucose transport.

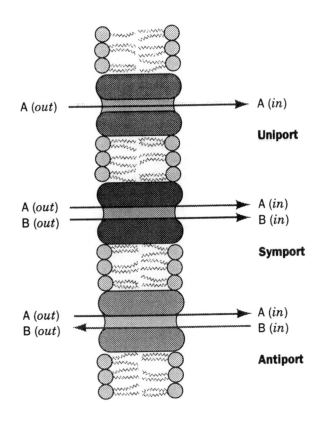

Figure 10-36. Uniport, symport, and antiport translocation systems.

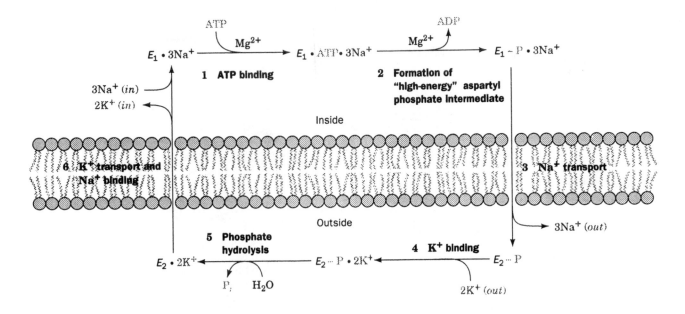

Figure 10-38. *Key to Function.* **Scheme for the active transport of Na$^+$ and K$^+$ by the (Na$^+$–K$^+$)–ATPase.**

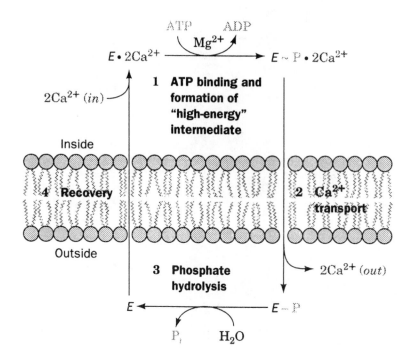

$$E \cdot 2Ca^{2+} \xrightarrow[\text{Mg}^{2+}]{\text{ATP} \quad \text{ADP}} E \sim P \cdot 2Ca^{2+}$$

$2Ca^{2+}$ (*in*)

1 ATP binding and formation of "high-energy" intermediate

Inside

4 Recovery

2 Ca^{2+} transport

Outside

3 Phosphate hydrolysis

$2Ca^{2+}$ (*out*)

$$E \longleftarrow E - P$$

$P_i \qquad H_2O$

Figure 10-39. Scheme for the active transport of Ca^{2+} by the Ca^{2+}–ATPase.

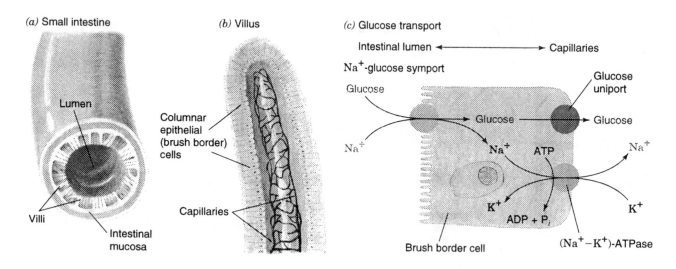

(*a*) Small intestine

Lumen

Villi

Intestinal mucosa

(*b*) Villus

Columnar epithelial (brush border) cells

Capillaries

(*c*) Glucose transport

Intestinal lumen ⟷ Capillaries

Na^+-glucose symport

Glucose

Na^+

Glucose

Glucose uniport

Glucose

Na^+

ATP

Na^+

K^+

ADP + P_i

K^+

Brush border cell

(Na^+-K^+)-ATPase

Figure 10-40. Glucose transport in the intestinal epithelium.

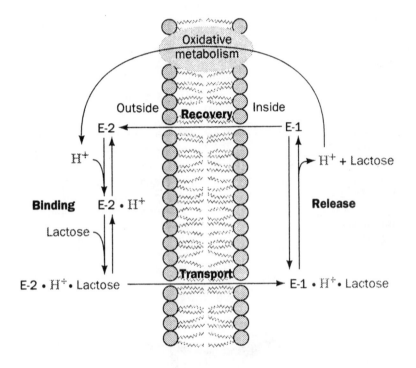

Figure 10-41. Scheme for the cotransport of H^+ and lactose by lactose permease in *E. coli.*

CHAPTER 11

ENZYMATIC CATALYSIS

Table 11-1. Catalytic Power of Some Enzymes

Enzyme	Nonenzymatic Reaction Rate (s^{-1})	Enzymatic Reaction Rate (s^{-1})	Rate Enhancement
Carbonic anhydrase	1.3×10^{-1}	1×10^{6}	7.7×10^{6}
Chorismate mutase	2.6×10^{-5}	50	1.9×10^{6}
Triose phosphate isomerase	4.3×10^{-6}	4300	1.0×10^{9}
Carboxypeptidase A	3.0×10^{-9}	578	1.9×10^{11}
AMP nucleosidase	1.0×10^{-11}	60	6.0×10^{12}
Staphylococcal nuclease	1.7×10^{-13}	95	5.6×10^{14}

Source: Radzicka, A. and Wolfenden, R., *Science* 267, 91 (1995).

Table 11-2. Enzyme Classification According to Reaction Type

Classification	Type of Reaction Catalyzed
1. Oxidoreductases	Oxidation–reduction reactions
2. Transferases	Transfer of functional groups
3. Hydrolases	Hydrolysis reactions
4. Lyases	Group elimination to form double bonds
5. Isomerases	Isomerization
6. Ligases	Bond formation coupled with ATP hydrolysis

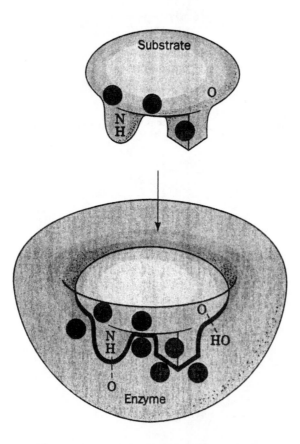

Figure 11-1. An enzyme–substrate complex.

DEMONSTRATION OF ENZYME STEREOSPECIFICITY

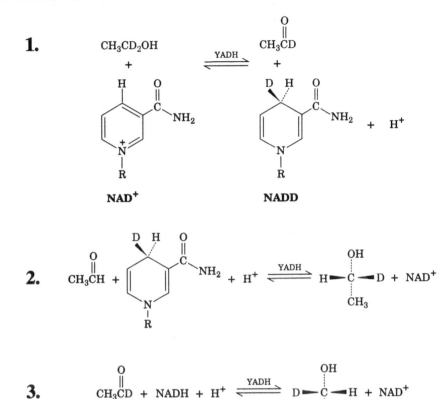

1.

CH_3CD_2OH + NAD⁺ $\underset{\text{YADH}}{\rightleftharpoons}$ CH_3CD (=O) + NADD + H⁺

NAD⁺   NADD

2. CH_3CH (=O) + NADD + H⁺ $\underset{\text{YADH}}{\rightleftharpoons}$ $H-C(OH)(CH_3)-D$ + NAD⁺

3. CH_3CD (=O) + NADH + H⁺ $\underset{\text{YADH}}{\rightleftharpoons}$ $D-C(OH)(CH_3)-H$ + NAD⁺

Table 11-3. **Characteristics of Common Coenzymes**

Coenzyme	Reaction Mediated	Vitamin Source	Human Deficiency Disease
Biocytin	Carboxylation	Biotin	a
Coenzyme A	Acyl transfer	Pantothenate	a
Cobalamin coenzymes	Alkylation	Cobalamin (B_{12})	Pernicious anemia
Flavin coenzymes	Oxidation–reduction	Riboflavin (B_2)	a
Lipoic acid	Acyl transfer	—	a
Nicotinamide coenzymes	Oxidation–reduction	Nicotinamide (niacin)	Pellagra
Pyridoxal phosphate	Amino group transfer	Pyridoxine (B_6)	a
Tetrahydrofolate	One-carbon group transfer	Folic acid	Megaloblastic anemia
Thiamine pyrophosphate	Aldehyde transfer	Thiamine (B_1)	Beriberi

[a]No specific name; deficiency in humans is rare or unobserved.

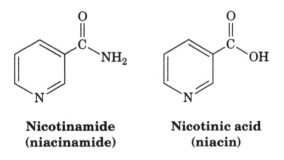

Nicotinamide
(niacinamide)   **Nicotinic acid**
(niacin)

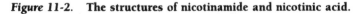

Figure 11-2. **The structures of nicotinamide and nicotinic acid.**

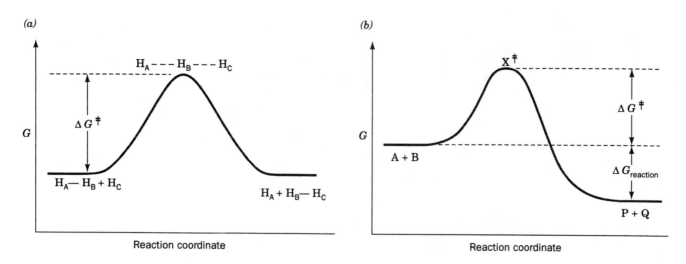

(a)

H_A - - - H_B - - - H_C

G

$\Delta G^{\ddagger}$

H_A — H_B + H_C

H_A + H_B — H_C

Reaction coordinate

(b)

$X^{\ddagger}$

G

A + B

$\Delta G^{\ddagger}$

$\Delta G_{reaction}$

P + Q

Reaction coordinate

Figure 11-3. **Transition state diagram.**

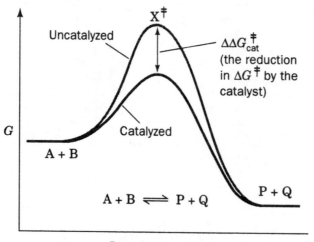

$X^{\ddagger}$

Uncatalyzed

$\Delta\Delta G_{cat}^{\ddagger}$
(the reduction
in $\Delta G^{\ddagger}$ by the
catalyst)

Catalyzed

G

A + B

A + B $\rightleftharpoons$ P + Q

P + Q

Reaction coordinate

Figure 11-5. **Effect of a catalyst on the transition state diagram of a reaction.**

TYPES OF CATALYTIC MECHANISMS

1. Acid–base catalysis
2. Covalent catalysis
3. Metal ion catalysis
4. Electrostatic catalysis
5. Proximity and orientation effects
6. Preferential binding of the transition state complex

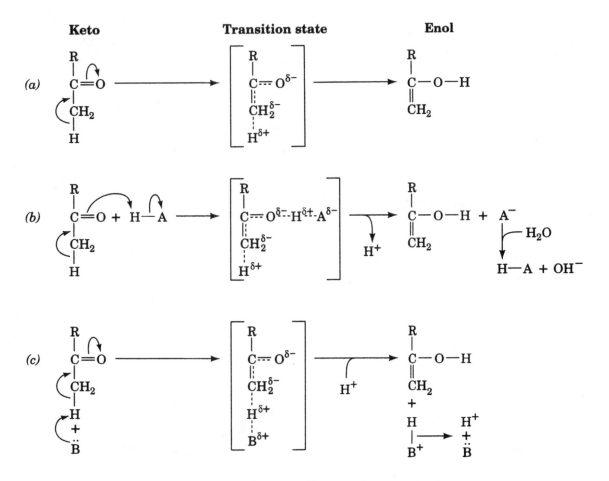

Figure 11-6. **Mechanisms of keto–enol tautomerization.**

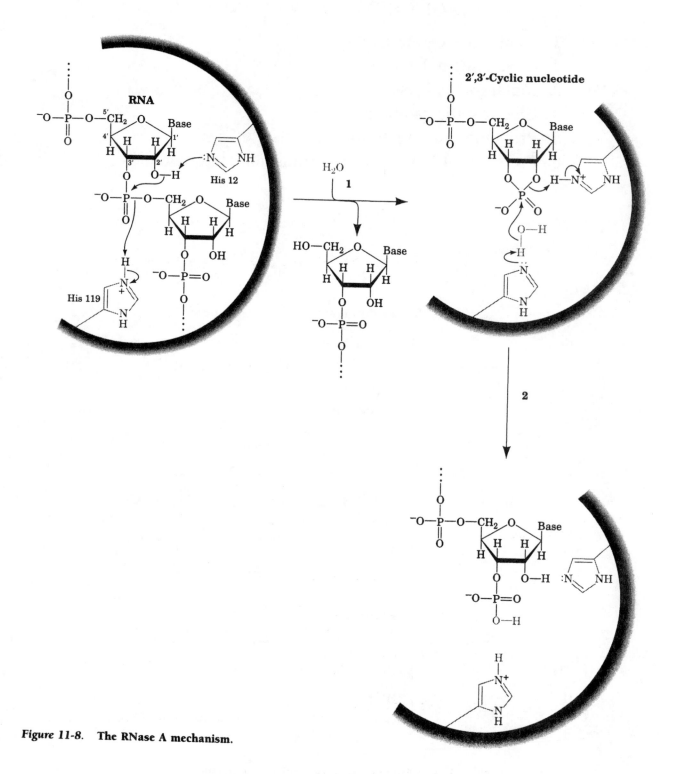

Figure 11-8. The RNase A mechanism.

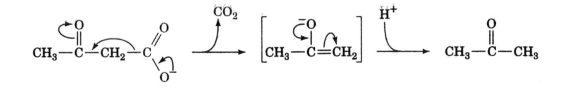

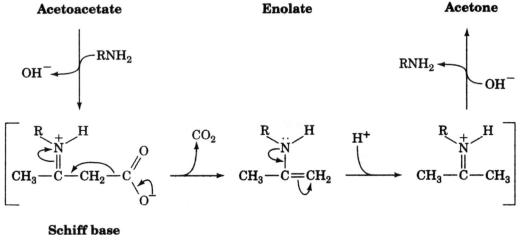

Acetoacetate **Enolate** **Acetone**

Schiff base
(imine)

Figure 11-9. **The decarboxylation of acetoacetate.**

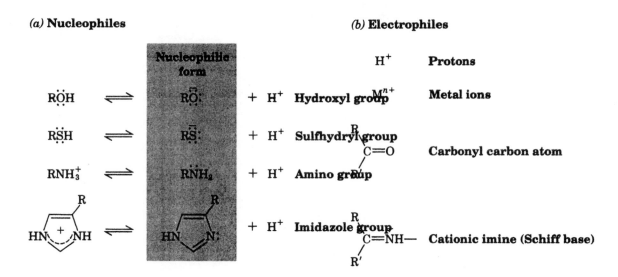

(a) **Nucleophiles**   *(b)* **Electrophiles**

			H^+	**Protons**
$R\ddot{O}H$	$\rightleftharpoons$	$R\ddot{\ddot{O}}:^-$	$+ H^+$ **Hydroxyl group** M^{n+}	**Metal ions**
$R\ddot{S}H$	$\rightleftharpoons$	$R\ddot{\ddot{S}}:^-$	$+ H^+$ **Sulfhydryl group**	
RNH_3^+	$\rightleftharpoons$	$R\ddot{N}H_2$	$+ H^+$ **Amino group**	**Carbonyl carbon atom**
				Cationic imine (Schiff base)

Figure 11-10. **Biologically important nucleophilic and electrophilic groups.**

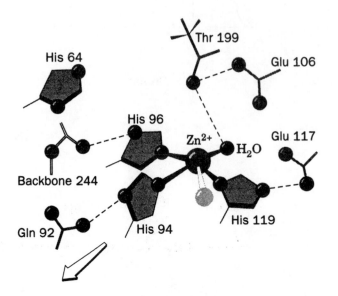

Figure 11-11. **The active site of human carbonic anhydrase.**

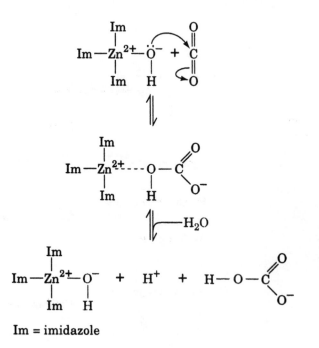

Im = imidazole

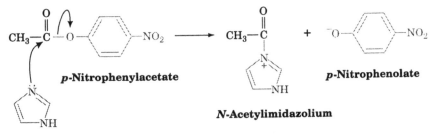

Imidazole

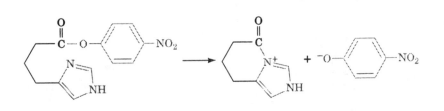

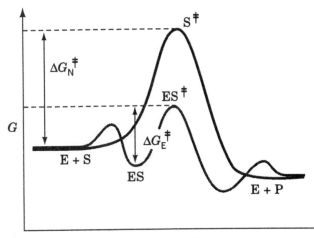

Figure 11-13. Key to Function. Effect of preferential transition state binding.

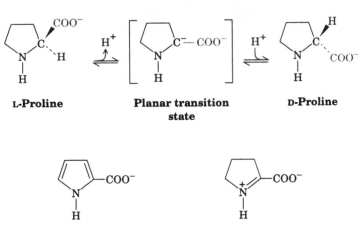

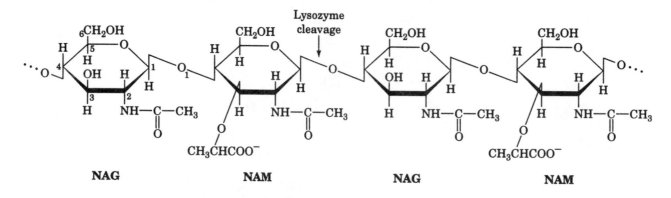

Figure 11-14. The lysozyme cleavage site.

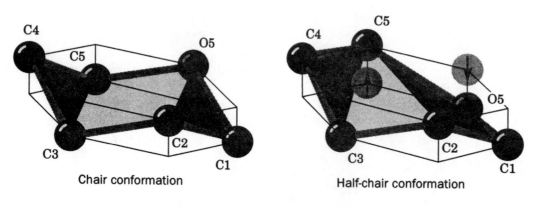

Chair conformation

Half-chair conformation

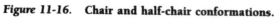

Figure 11-16. Chair and half-chair conformations.

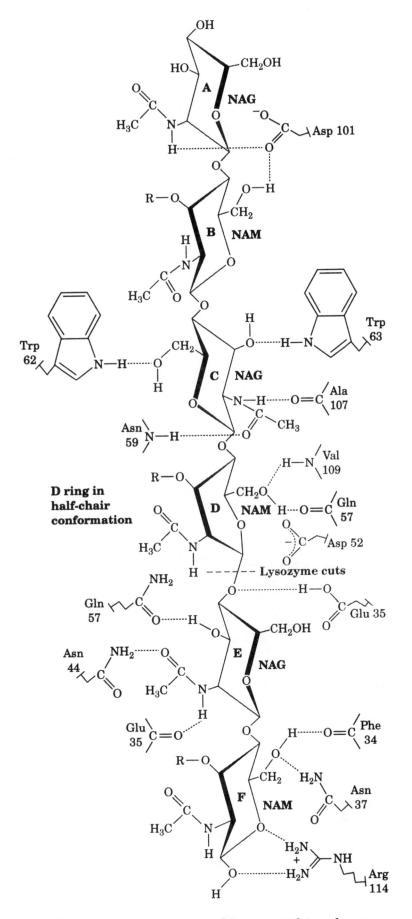

Figure 11-17. The interactions of lysozyme with its substrate.

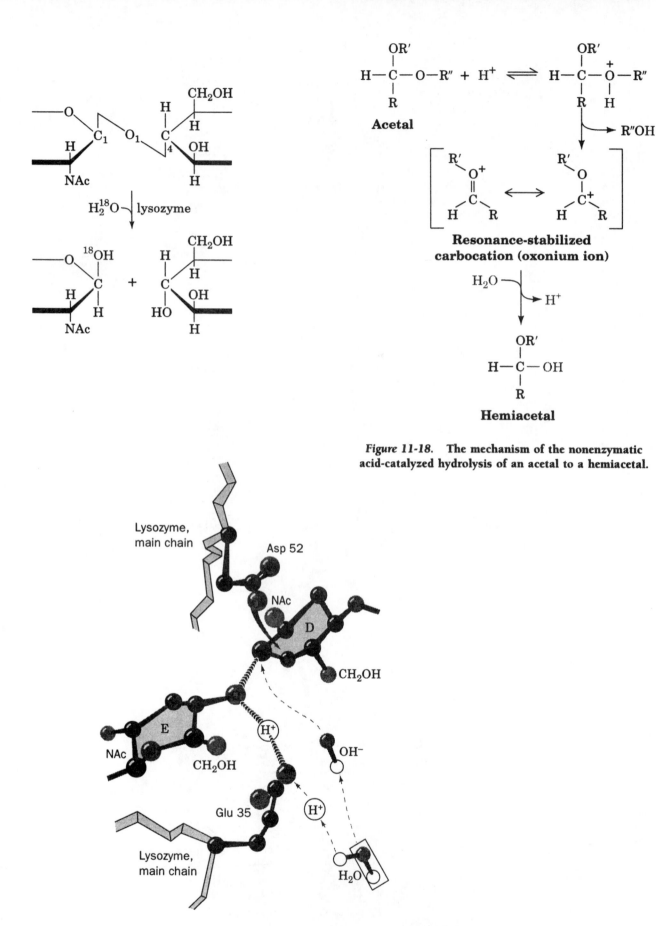

Figure 11-18. The mechanism of the nonenzymatic acid-catalyzed hydrolysis of an acetal to a hemiacetal.

Figure 11-19. *Key to Function.* The Phillips mechanism for the lysozyme reaction.

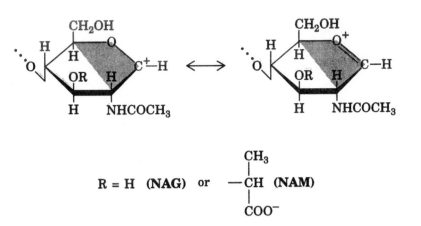

Figure 11-20. Transition state for the lysozyme reaction.

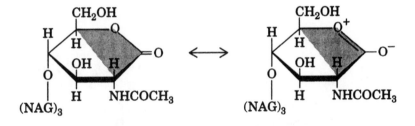

Figure 11-21. The δ-lactone analog of (NAG)$_4$.

Table 11-4. Binding Free Energies of HEW Lysozyme Subsites

Site	Bound Saccharide	Binding Free Energy $(kJ \cdot mol^{-1})$
A	NAG	−7.5
B	NAM	−12.3
C	NAG	−23.8
D	NAM	+12.1
E	NAG	−7.1
F	NAM	−7.1

Source: Chipman, D.M. and Sharon, N., *Science* 165, 459 (1969).

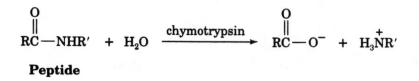

Peptide

Ester

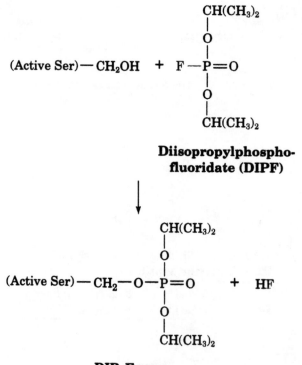

**Diisopropylphospho-
fluoridate (DIPF)**

DIP–Enzyme

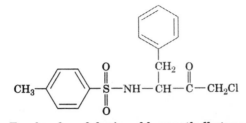

Tosyl-L-phenylalanine chloromethylketone

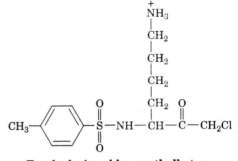

Tosyl-L-lysine chloromethylketone

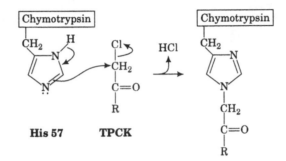

His 57   **TPCK**

Figure 11-22.   **Reaction of TPCK with His 57 of chymotrypsin.**

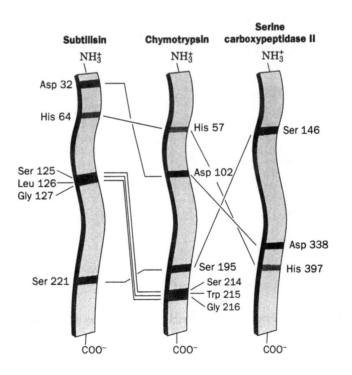

Figure 11-25.   **A diagram indicating the relative positions of the active site residues of three unrelated serine proteases.**

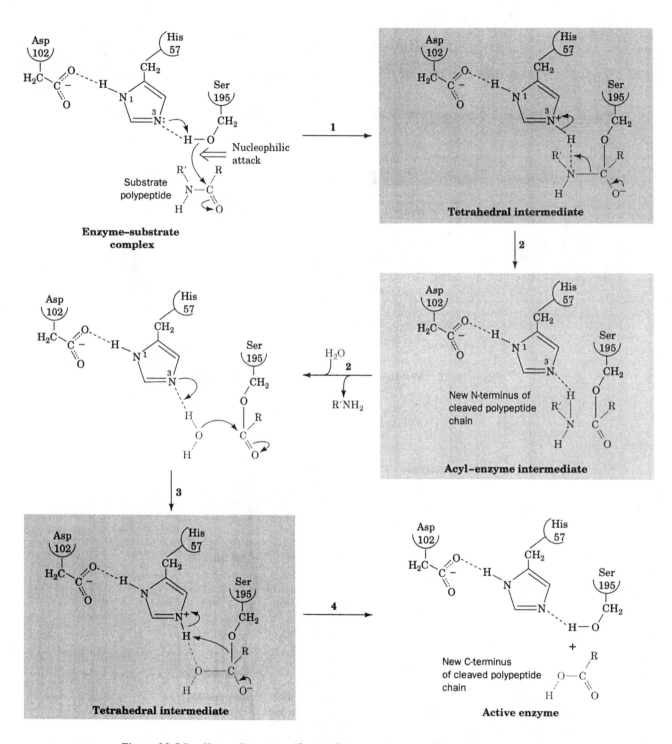

Figure 11-26. *Key to Function.* The catalytic mechanism of the serine proteases.

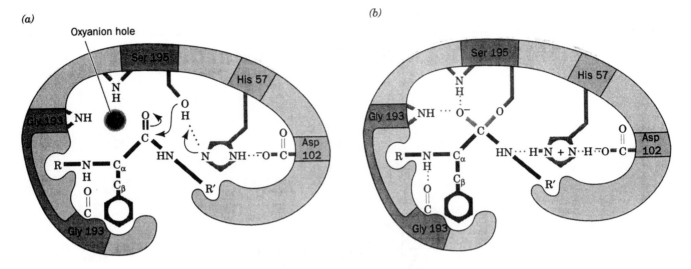

Figure 11-27. Transition state stabilization in the serine proteases.

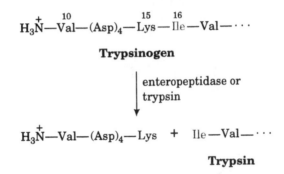

Figure 11-29. The activation of trypsinogen to trypsin.

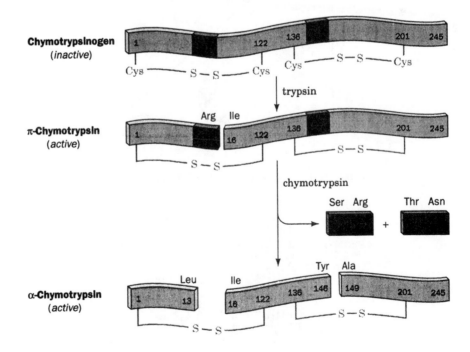

Figure 11-30. The activation of chymotrypsinogen by proteolytic cleavage.

CHAPTER 12

ENZYME KINETICS, INHIBITION, AND REGULATION

$$A \longrightarrow P$$

$$v = \frac{d[P]}{dt} = -\frac{d[A]}{dt} = k[A] \qquad [12\text{-}1]$$

$$\frac{d[A]}{[A]} = d \ln[A] = -k \, dt \qquad [12\text{-}4]$$

$$\int_{[A]_o}^{[A]} d \ln[A] = -k \int_0^t dt \qquad [12\text{-}5]$$

$$\boxed{\ln[A] = \ln[A]_o - kt} \qquad [12\text{-}6]$$

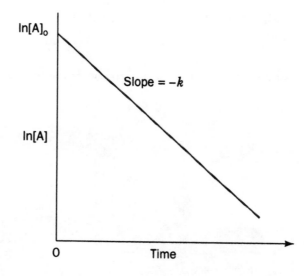

Figure 12-1.   A plot of a first-order rate equation.

$$\ln\left(\frac{[A]_o/2}{[A]_o}\right) = -kt_{1/2} \qquad [12\text{-}8]$$

$$t_{1/2} = \frac{\ln 2}{k} = \frac{0.693}{k} \qquad [12\text{-}9]$$

$$E + S \underset{k_{-1}}{\overset{k_1}{\rightleftharpoons}} ES \xrightarrow{k_2} P + E \qquad [12\text{-}12]$$

$$v = \frac{d[\text{P}]}{dt} = k_2[\text{ES}] \qquad [12\text{-}13]$$

$$\frac{d[\text{ES}]}{dt} = k_1[\text{E}][\text{S}] - k_{-1}[\text{ES}] - k_2[\text{ES}] \qquad [12\text{-}14]$$

$$K_\text{S} = \frac{k_{-1}}{k_1} = \frac{[\text{E}][\text{S}]}{[\text{ES}]} \qquad [12\text{-}15]$$

$$\frac{d[\text{ES}]}{dt} = 0 \qquad [12\text{-}16]$$

$$[\text{E}]_\text{T} = [\text{E}] + [\text{ES}] \qquad [12\text{-}17]$$

$$k_1[\text{E}][\text{S}] = k_{-1}[\text{ES}] + k_2[\text{ES}] \qquad [12\text{-}18]$$

$$\frac{([\text{E}]_\text{T} - [\text{ES}])[\text{S}]}{[\text{ES}]} = \frac{k_{-1} + k_2}{k_1} \qquad [12\text{-}19]$$

$$K_M = \frac{k_{-1} + k_2}{k_1} \qquad [12\text{-}20]$$

$$K_M[\text{ES}] = ([\text{E}]_\text{T} - [\text{ES}])[\text{S}] \qquad [12\text{-}21]$$

$$[\text{ES}] = \frac{[\text{E}]_\text{T}[\text{S}]}{K_M + [\text{S}]} \qquad [12\text{-}22]$$

$$v_o = \left(\frac{d[P]}{dt}\right)_{t=0} = k_2[ES] = \frac{k_2[E]_T[S]}{K_M + [S]} \qquad [12\text{-}23]$$

$$V_{max} = k_2[E]_T \qquad [12\text{-}24]$$

$$v_o = \frac{V_{max}[S]}{K_M + [S]} \qquad [12\text{-}25]$$

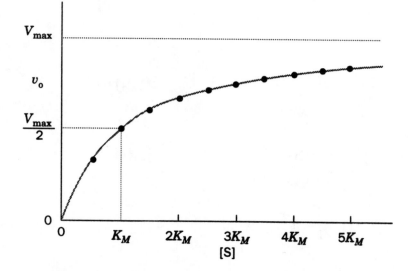
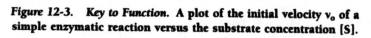

Figure 12-3. *Key to Function.* A plot of the initial velocity v_o of a simple enzymatic reaction versus the substrate concentration [S].

$$K_M = \frac{k_{-1}}{k_1} + \frac{k_2}{k_1} = K_S + \frac{k_2}{k_1} \qquad [12\text{-}26]$$

$$k_{cat} = \frac{V_{max}}{[E]_T} \qquad [12\text{-}27]$$

$$v_o \approx \left(\frac{k_2}{K_M}\right)[E]_T[S] \approx \left(\frac{k_{cat}}{K_M}\right)[E][S] \qquad [12\text{-}28]$$

$$\frac{1}{v_o} = \left(\frac{K_M}{V_{max}}\right)\frac{1}{[S]} + \frac{1}{V_{max}} \qquad \text{[12-29]}$$

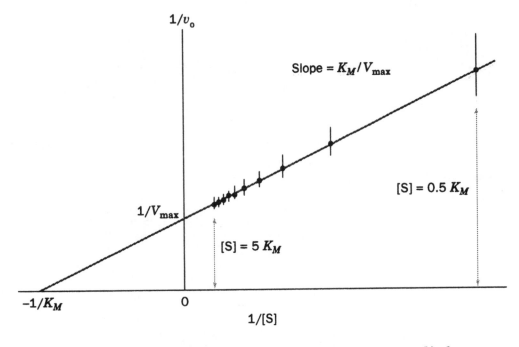

Figure 12-4. **Key to Function. A double-reciprocal (Lineweaver–Burk) plot.**

(a)

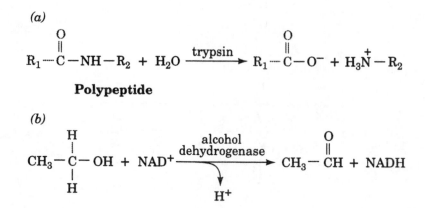

Polypeptide

(b)

Figure 12-5. **Some bisubstrate reactions.**

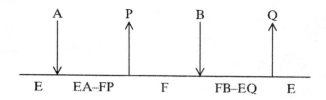

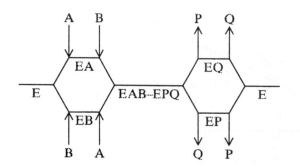

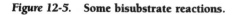

$$E + S \xrightleftharpoons[k_{-1}]{k_1} ES \xrightarrow{k_2} P + E$$

$$+$$

$$I$$

$$K_I \Updownarrow$$

$$EI + S \longrightarrow \text{NO REACTION}$$

$$K_I = \frac{[E][I]}{[EI]} \qquad\qquad [12\text{-}30]$$

$$v_o = \frac{V_{max}[S]}{\alpha K_M + [S]} \qquad\qquad [12\text{-}31]$$

$$\alpha = 1 + \frac{[I]}{K_I} \qquad\qquad [12\text{-}32]$$

$$\frac{1}{v_o} = \left(\frac{\alpha K_M}{V_{max}}\right)\frac{1}{[S]} + \frac{1}{V_{max}} \qquad\qquad [12\text{-}33]$$

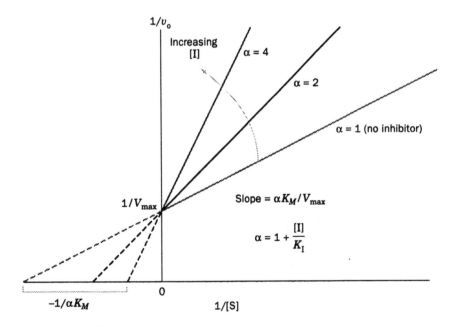

Figure 12-7. A Lineweaver–Burk plot of the competitively inhibited Michaelis–Menten enzyme described by Fig. 12-6.

HIV PROTEASE INHIBITORS

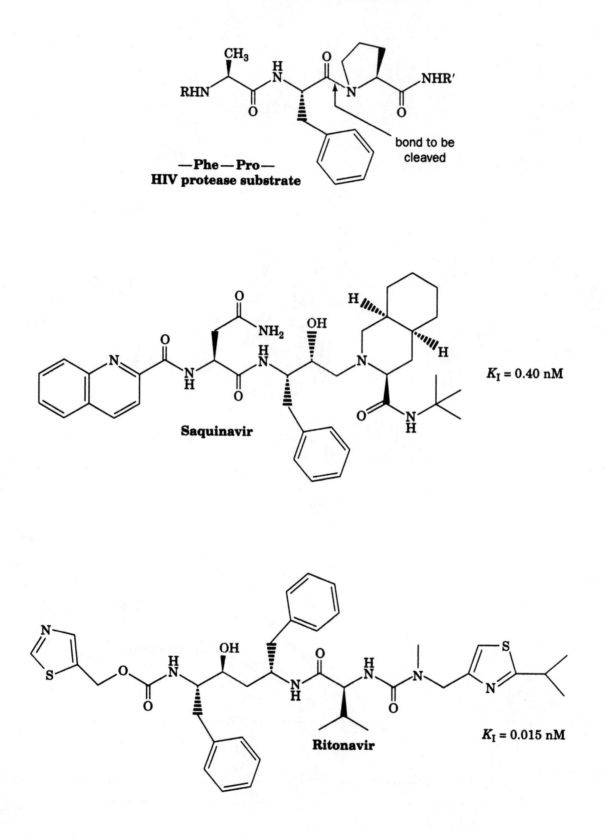

—Phe—Pro—
HIV protease substrate

bond to be cleaved

Saquinavir

$K_I = 0.40$ nM

Ritonavir

$K_I = 0.015$ nM

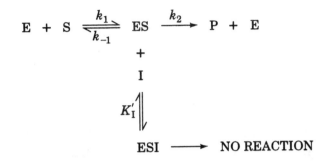

$$K_I' = \frac{[\text{ES}][\text{I}]}{[\text{ESI}]} \qquad\qquad [12\text{-}34]$$

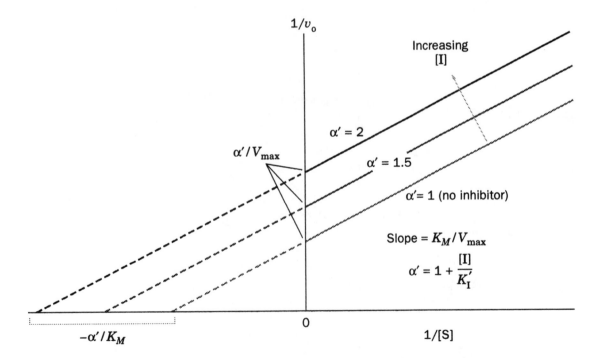

Figure 12-8. **A Lineweaver–Burk plot of a Michaelis–Menten enzyme in the presence of an uncompetitive inhibitor.**

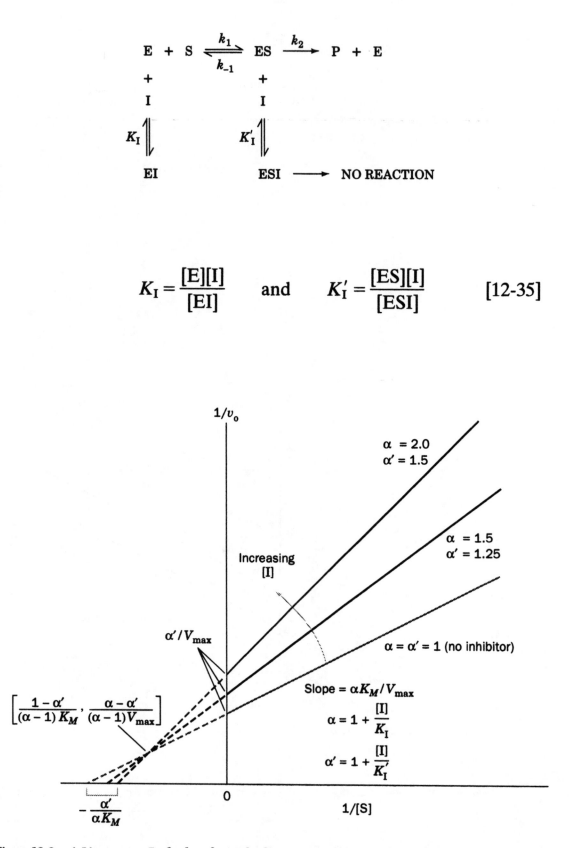

$$E + S \underset{k_{-1}}{\overset{k_1}{\rightleftharpoons}} ES \xrightarrow{k_2} P + E$$

$$
\begin{array}{ccc}
+ & & + \\
I & & I
\end{array}
$$

$$K_I \updownarrow \qquad K_I' \updownarrow$$

$$EI \qquad\qquad ESI \longrightarrow NO\ REACTION$$

$$K_I = \frac{[E][I]}{[EI]} \qquad \text{and} \qquad K_I' = \frac{[ES][I]}{[ESI]} \qquad\qquad [12\text{-}35]$$

1/v_o

$\alpha = 2.0$
$\alpha' = 1.5$

$\alpha = 1.5$
$\alpha' = 1.25$

Increasing
[I]

α'/V_{max}

$\alpha = \alpha' = 1$ (no inhibitor)

Slope = $\alpha K_M/V_{max}$

$\alpha = 1 + \dfrac{[I]}{K_I}$

$\alpha' = 1 + \dfrac{[I]}{K_I'}$

$\left[\dfrac{1 - \alpha'}{(\alpha - 1)\,K_M},\ \dfrac{\alpha - \alpha'}{(\alpha - 1)\,V_{max}} \right]$

$-\dfrac{\alpha'}{\alpha K_M}$

0

1/[S]

Figure 12-9. A Lineweaver–Burk plot of a Michaelis–Menten enzyme in the presence of a mixed inhibitor.

Table 12-2. **Effects of Inhibitors on Michaelis–Menten Reactions**[a]

Type of Inhibition	Michaelis–Menten Equation	Lineweaver–Burk Equation	Effect of Inhibitor
None	$v_o = \dfrac{V_{max}[S]}{K_M + [S]}$	$\dfrac{1}{v_o} = \dfrac{K_M}{V_{max}} \dfrac{1}{[S]} + \dfrac{1}{V_{max}}$	None
Competitive	$v_o = \dfrac{V_{max}[S]}{\alpha K_M + [S]}$	$\dfrac{1}{v_o} = \dfrac{\alpha K_M}{V_{max}} \dfrac{1}{[S]} + \dfrac{1}{V_{max}}$	Increases K_M
Uncompetitive	$v_o = \dfrac{V_{max}[S]}{K_M + \alpha'[S]}$	$\dfrac{1}{v_o} = \dfrac{K_M}{V_{max}} \dfrac{1}{[S]} + \dfrac{\alpha'}{V_{max}}$	Decreases K_M and V_{max}
Mixed (noncompetitive)	$v_o = \dfrac{V_{max}[S]}{\alpha K_M + \alpha'[S]}$	$\dfrac{1}{v_o} = \dfrac{\alpha K_M}{V_{max}} \dfrac{1}{[S]} + \dfrac{\alpha'}{V_{max}}$	Decreases V_{max}; may increase or decrease K_M

[a] $\alpha = 1 + \dfrac{[I]}{K_I}$ and $\alpha' = 1 + \dfrac{[I]}{K_I'}$

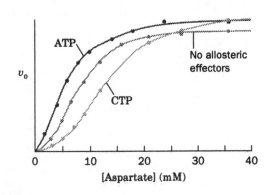

Figure 12-10. Plot of v_o versus [Aspartate] for the ATCase reaction.

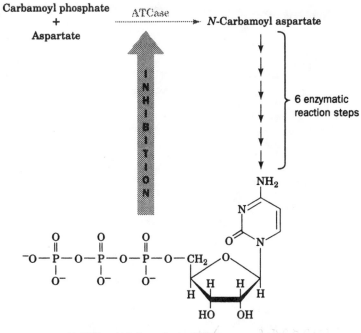

Figure 12-11. A schematic representation of the pyrimidine biosynthesis pathway.

CHAPTER 13

INTRODUCTION TO METABOLISM

2 phases ← *break down*

Catabolism + Anabolism ← *make things.*

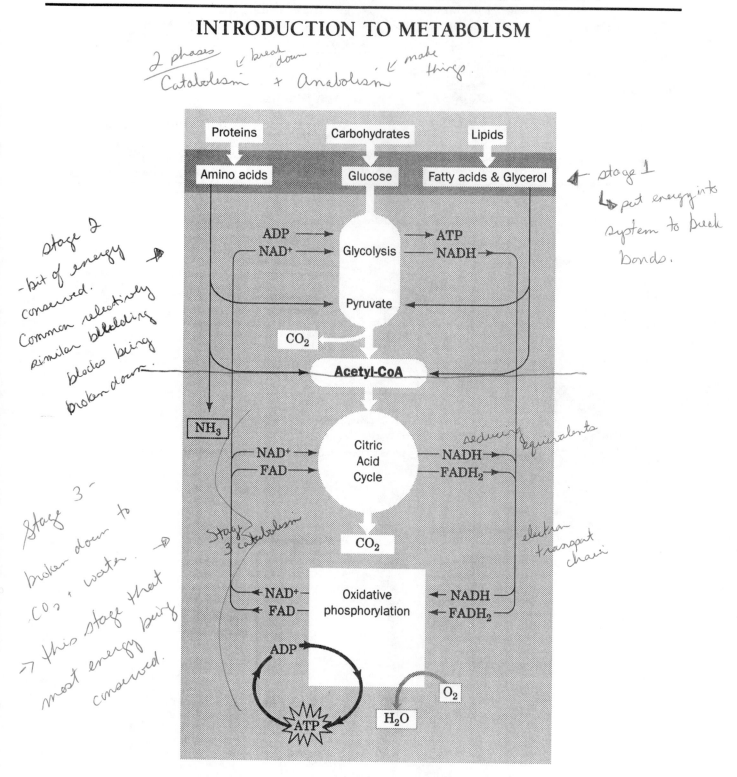

stage 1
↳ *put energy into system to break bonds.*

stage 2
- bit of energy conserved.
Common relatively similar building blocks being broken down.

stage 3 -
broken down to CO₂ + water.
→ this stage that most energy being conserved.

Stage 3 catabolism

reducing equivalents

electron transport chain

Figure 13-2. *Key to Metabolism.* **Overview of catabolism.**

converge \/

Table 13-1. Metabolic Functions of Eukaryotic Organelles

Organelle	Function
Mitochondrion	Citric acid cycle, oxidative phosphorylation, fatty acid oxidation, amino acid breakdown
Cytosol	Glycolysis, pentose phosphate pathway, fatty acid biosynthesis, many reactions of gluconeogenesis
Lysosomes	Enzymatic digestion of cell components and ingested matter
Nucleus	DNA replication and transcription, RNA processing
Golgi apparatus	Posttranslational processing of membrane and secretory proteins; formation of plasma membrane and secretory vesicles
Rough endoplasmic reticulum	Synthesis of membrane-bound and secretory proteins
Smooth endoplasmic reticulum	Lipid and steroid biosynthesis
Peroxisomes (glyoxysomes in plants)	Oxidative reactions catalyzed by amino acid oxidases and catalase; glyoxylate cycle reactions in plants

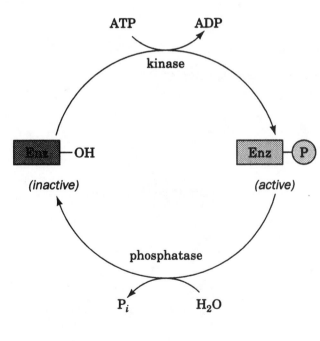

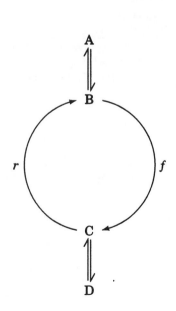

135

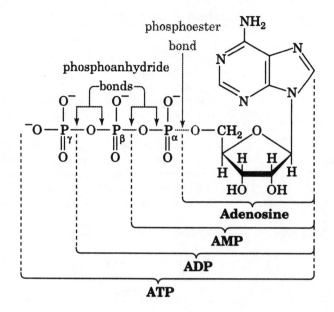

Figure 13-3. *Key to Structure.* The structure of ATP indicating its relationship to ADP, AMP, and adenosine.

Table 13-2. Standard Free Energies of Phosphate Hydrolysis of Some Compounds of Biological Interest

Compound	$\Delta G^{o\prime}$ $(kJ \cdot mol^{-1})$
Phosphoenolpyruvate	−61.9
1,3-Bisphosphoglycerate	−49.4
Acetyl phosphate	−43.1
Phosphocreatine	−43.1
PP_i	−33.5
ATP ($\rightarrow$ AMP + PP_i)	−32.2
ATP ($\rightarrow$ ADP + P_i)	−30.5
Glucose-1-phosphate	−20.9
Fructose-6-phosphate	−13.8
Glucose-6-phosphate	−13.8
Glycerol-3-phosphate	−9.2

Source: Jencks, W.P., in Fasman, G.D. (Ed.), *Handbook of Biochemistry and Molecular Biology* (3rd ed.), Physical and Chemical Data, Vol. I, pp. 296–304, CRC Press (1976).

handwritten annotations:

Really not hydrolysis that occurs, just use the H₂O. It's actually the transfer of PO₄.

just gives idea of energy.

middle

2-3x more energy than AMP

AMP + H₂O ——→ ~3.0 Kcal/mol

ATP in middle of table.

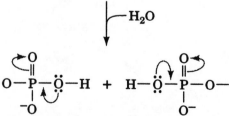

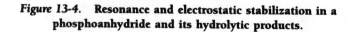

Figure 13-4. **Resonance and electrostatic stabilization in a phosphoanhydride and its hydrolytic products.**

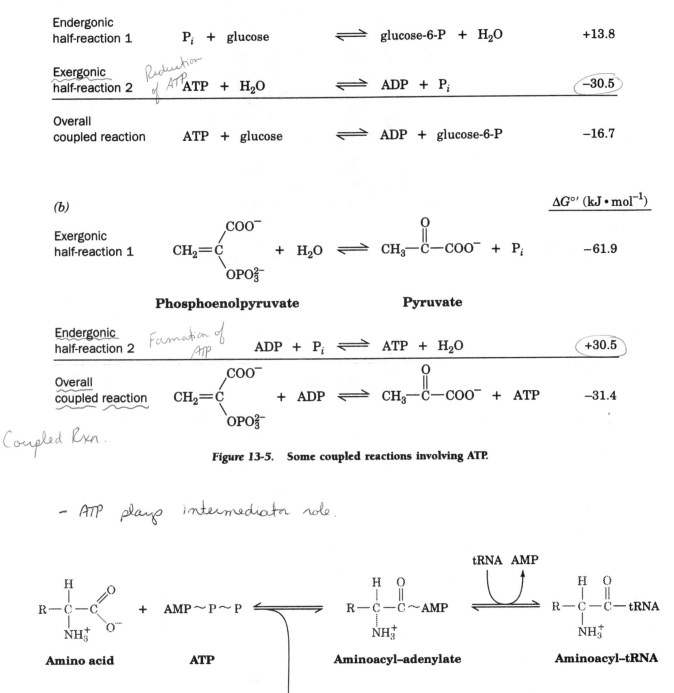

			$\Delta G^{\circ\prime}$ (kJ · mol^{-1})

(a)

Endergonic half-reaction 1 $\quad$ P$_i$ + glucose $\rightleftharpoons$ glucose-6-P + H$_2$O $\quad$ +13.8

Exergonic half-reaction 2 $\quad$ *Reduction of ATP* ATP + H$_2$O $\rightleftharpoons$ ADP + P$_i$ $\quad$ −30.5

Overall coupled reaction $\quad$ ATP + glucose $\rightleftharpoons$ ADP + glucose-6-P $\quad$ −16.7

(b)

Exergonic half-reaction 1

Phosphoenolpyruvate $\qquad$ **Pyruvate** $\qquad$ −61.9

Endergonic half-reaction 2 $\quad$ *Formation of ATP* $\quad$ ADP + P$_i$ $\rightleftharpoons$ ATP + H$_2$O $\quad$ +30.5

Overall coupled reaction $\qquad$ −31.4

Figure 13-5. Some coupled reactions involving ATP.

Coupled Rxn.

− ATP plays intermediate role.

Amino acid $\qquad$ **ATP** $\qquad$ **Aminoacyl–adenylate** $\qquad$ **Aminoacyl–tRNA**

Figure 13-6. Pyrophosphate cleavage in the synthesis of an aminoacyl–tRNA.

137

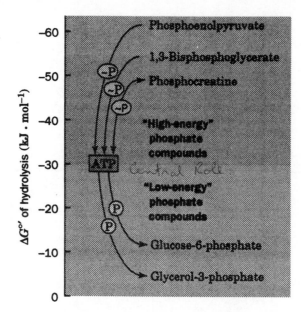

Figure 13-7. Position of ATP relative to "high-energy" and "low-energy" phosphate compounds.

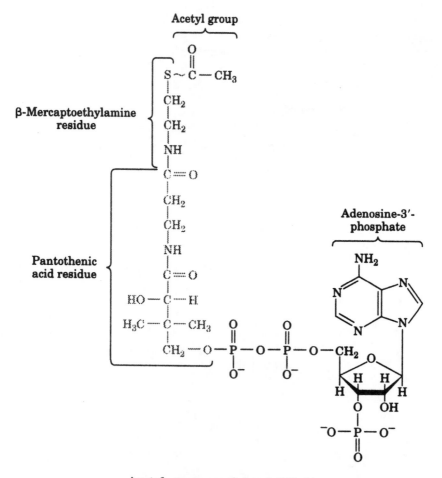

Acetyl-coenzyme A (acetyl-CoA)

Figure 13-9. The chemical structure of acetyl-CoA.

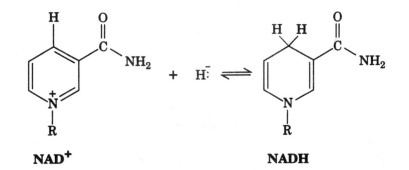

NAD⁺   NADH

Figure 13-10. **Reduction of NAD⁺ to NADH.**

Flavin adenine dinucleotide (FAD)
(oxidized or quinone form)

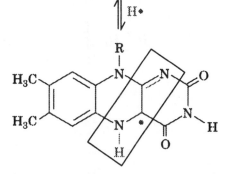

FADH· (radical or semiquinone form)

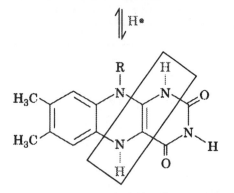

FADH₂ (reduced or hydroquinone form)

Figure 13-11. **Reduction of FAD to FADH₂.**

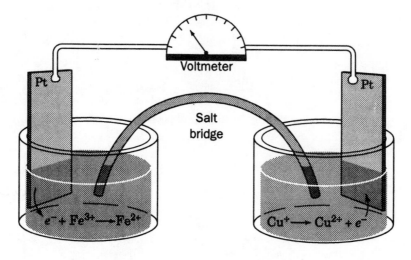

Figure 13-12. An electrochemical cell.

$$\Delta G = \Delta G^\circ + RT \ln \left(\frac{[A_{red}][B_{ox}^{n+}]}{[A_{ox}^{n+}][B_{red}]} \right) \qquad [13\text{-}3]$$

$$\Delta G = -n\mathscr{F}\Delta\mathscr{E} \qquad [13\text{-}6]$$

$$\boxed{\Delta\mathscr{E} = \Delta\mathscr{E}^\circ - \frac{RT}{n\mathscr{F}} \ln \left(\frac{[A_{red}][B_{ox}^{n+}]}{[A_{ox}^{n+}][B_{red}]} \right)} \qquad [13\text{-}7]$$

$$\Delta\mathscr{E}^\circ = \mathscr{E}^\circ_{(e^-\text{acceptor})} - \mathscr{E}^\circ_{(e^-\text{donor})} \qquad [13\text{-}10]$$

Table 13-3. Standard Reduction Potentials of Some Biochemically Important Half-Reactions

Half-Reaction	$\mathscr{E}\mathscr{E}$ (V)
$\frac{1}{2} O_2 + 2 H^+ + 2 e^- \rightleftharpoons H_2O$	0.815
$SO_4^{2-} + 2 H^+ + 2 e^- \rightleftharpoons SO_3^{2-} + H_2O$	0.48
$NO_3^- + 2 H^+ + 2 e^- \rightleftharpoons NO_2^- + H_2O$	0.42
Cytochrome a_3 (Fe^{3+}) $+ e^- \rightleftharpoons$ cytochrome a_3 (Fe^{2+})	0.385
$O_2(g) + 2 H^+ + 2 e^- \rightleftharpoons H_2O_2$	0.295
Cytochrome a (Fe^{3+}) $+ e^- \rightleftharpoons$ cytochrome a (Fe^{2+})	0.29
Cytochrome c (Fe^{3+}) $+ e^- \rightleftharpoons$ cytochrome c (Fe^{2+})	0.235
Cytochrome c_1 (Fe^{3+}) $+ e^- \rightleftharpoons$ cytochrome c_1 (Fe^{2+})	0.22
Cytochrome b (Fe^{3+}) $+ e^- \rightleftharpoons$ cytochrome b (Fe^{2+}) (*mitochondrial*)	0.077
Ubiquinone $+ 2 H^+ + 2 e^- \rightleftharpoons$ ubiquinol	0.045
Fumarate$^- + 2 H^+ + 2 e^- \rightleftharpoons$ succinate$^-$	0.031
FAD $+ 2 H^+ + 2 e^- \rightleftharpoons FADH_2$ (*in flavoproteins*)	~0.
Oxaloacetate$^- + 2 H^+ + 2 e^- \rightleftharpoons$ malate$^-$	−0.166
Pyruvate$^- + 2 H^+ + 2 e^- \rightleftharpoons$ lactate$^-$	−0.185
Acetaldehyde $+ 2 H^+ + 2 e^- \rightleftharpoons$ ethanol	−0.197
FAD $+ 2 H^+ + 2 e^- \rightleftharpoons FADH_2$ (*free coenzyme*)	−0.219
$S + 2 H^+ + 2 e^- \rightleftharpoons H_2S$	−0.23
Lipoic acid $+ 2 H^+ + 2 e^- \rightleftharpoons$ dihydrolipoic acid	−0.29
NAD$^+ + H^+ + 2 e^- \rightleftharpoons$ NADH	−0.315
NADP$^+ + H^+ + 2 e^- \rightleftharpoons$ NADPH	−0.320
Cystine $+ 2 H^+ + 2 e^- \rightleftharpoons$ 2 cysteine	−0.340
Acetoacetate$^- + 2 H^+ + 2 e^- \rightleftharpoons$ β-hydroxybutyrate$^-$	−0.346
$H^+ + e^- \rightleftharpoons \frac{1}{2} H_2$	−0.421
Acetate$^- + 3 H^+ + 2 e^- \rightleftharpoons$ acetaldehyde $+ H_2O$	−0.581

Source: Mostly from Loach, P.A., *In* Fasman, G.D. (Ed.), *Handbook of Biochemistry and Molecular Biology* (3rd ed.), Physical and Chemical Data, Vol. I, pp. 123–130, CRC Press (1976).

CHAPTER 14

GLUCOSE CATABOLISM

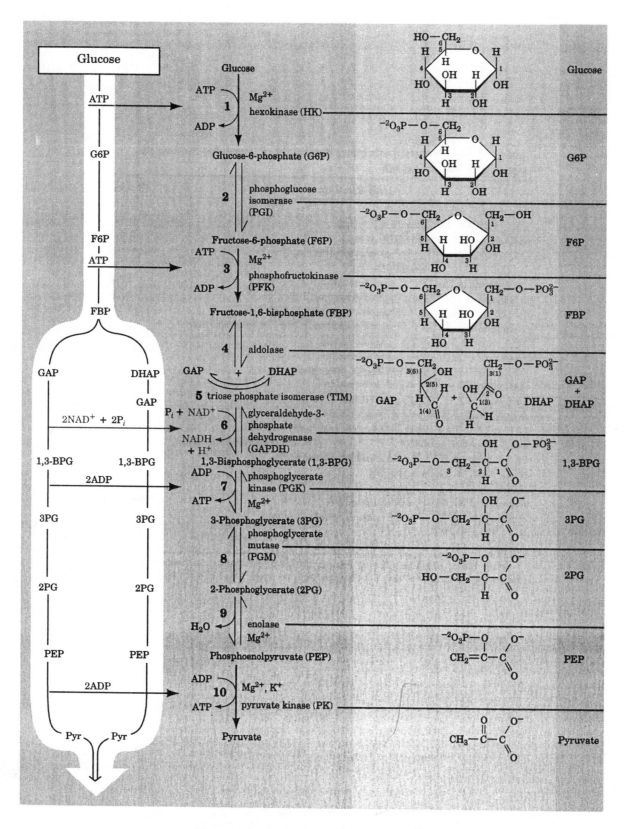

Figure 14-1. *Key to Metabolism. Glycolysis.*

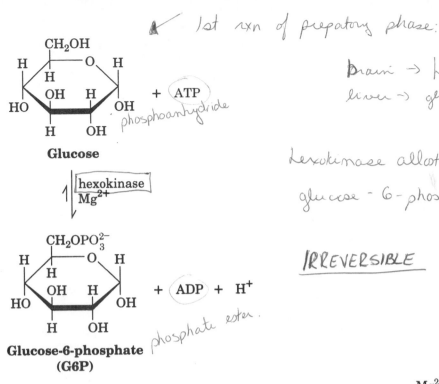

Glucose

Glucose-6-phosphate (G6P) *phosphate ester.*

hexokinase
Mg^{2+}

+ ATP *phosphoanhydride*

+ ADP + H^+

1st rxn of prepatory phase:

brain → hexokinase
liver → glucokinase

hexokinase allosterically inhibited by glucose-6-phosphate.

IRREVERSIBLE

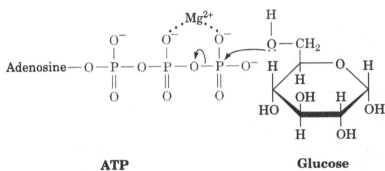

ATP   **Glucose**

Figure 14-2. **Substrate-induced conformational changes in yeast hexokinase.**

(a)   (b)

143

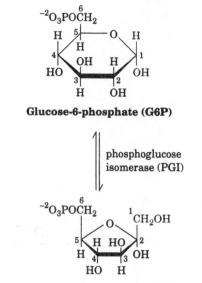

Glucose-6-phosphate (G6P)

phosphoglucose
isomerase (PGI)

Fructose-6-phosphate (F6P)

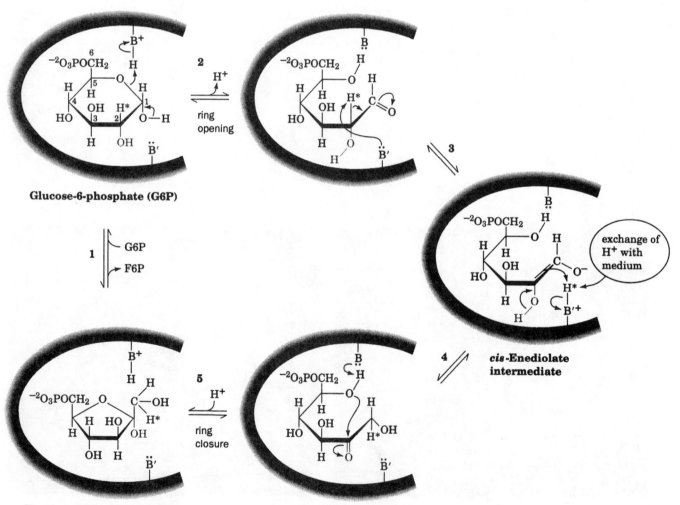

Figure 14-3. The reaction mechanism of phosphoglucose isomerase.

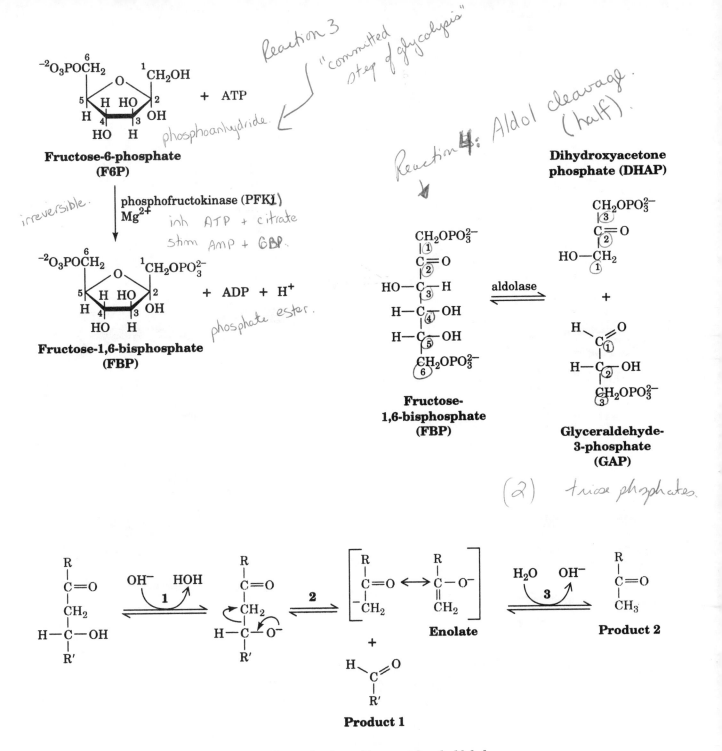

**Fructose-6-phosphate
(F6P)**

Reaction 3 "committed step of glycolysis"

phosphoanhydride.

irreversible.

phosphofructokinase (PFK1)
Mg^{2+}

inh ATP + citrate
stim AMP + GBP.

**Fructose-1,6-bisphosphate
(FBP)**

phosphate ester.

+ ATP

Reaction 4: Aldol cleavage. (half).

**Dihydroxyacetone
phosphate (DHAP)**

aldolase

**Fructose-
1,6-bisphosphate
(FBP)**

+ ADP + H^+

**Glyceraldehyde-
3-phosphate
(GAP)**

(2) triose phosphates.

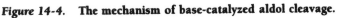

Enolate

Product 2

Product 1

Figure 14-4. **The mechanism of base-catalyzed aldol cleavage.**

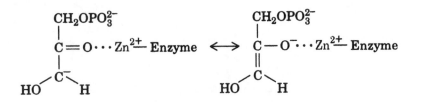

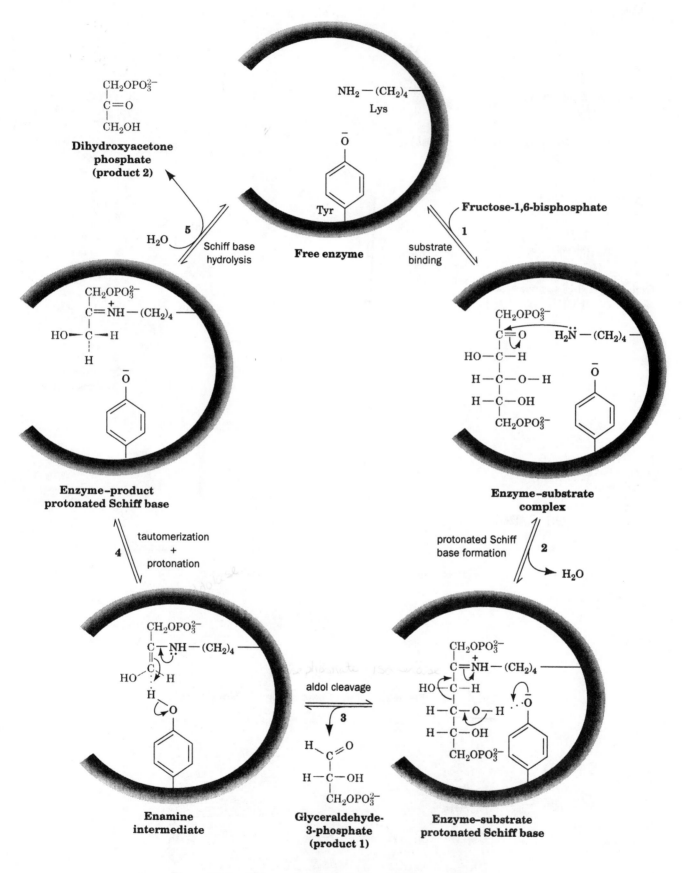

Figure 14-5. The enzymatic mechanism of Class I aldolase.

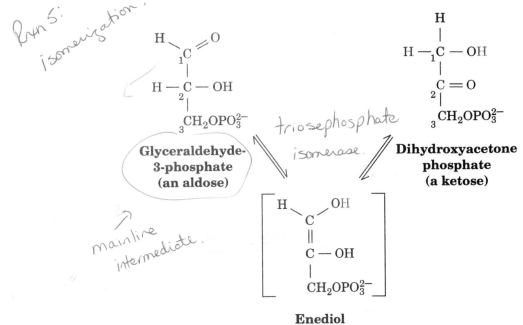

Rxn 5: isomerization.

H—C=0 (1)
H—C—OH (2)
CH$_2$OPO$_3^{2-}$ (3)

Glyceraldehyde-3-phosphate (an aldose)

mainline intermediate.

triosephosphate isomerase.

H (1)
H—C—OH
C=O (2)
CH$_2$OPO$_3^{2-}$ (3)

Dihydroxyacetone phosphate (a ketose)

$$\begin{bmatrix} \text{H}-\underset{|}{\overset{\text{OH}}{\text{C}}} \\ \text{C}-\text{OH} \\ | \\ \text{CH}_2\text{OPO}_3^{2-} \end{bmatrix}$$

Enediol intermediate

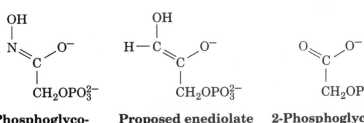

OH
N=C—O$^-$
CH$_2$OPO$_3^{2-}$

Phosphoglyco-hydroxamate

OH
H—C=C—O$^-$
CH$_2$OPO$_3^{2-}$

Proposed enediolate intermediate

O=C—O$^-$
CH$_2$OPO$_3^{2-}$

2-Phosphoglycolate

Figure 14-6. A ribbon diagram of yeast TIM in complex with its transition state analog 2-phosphoglycolate.

Preparatory Stage:

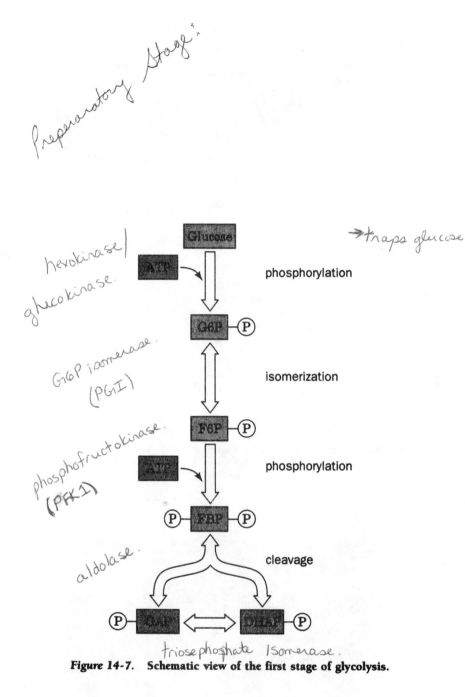

hexokinase/
glucokinase.

→ traps glucose

phosphorylation

G6P isomerase.
(PGI)

isomerization

phosphofructokinase.
(PFK1)

phosphorylation

aldolase.

cleavage

triosephosphate isomerase.

Figure 14-7. **Schematic view of the first stage of glycolysis.**

Energy Conservative Steps   #1 = #6.

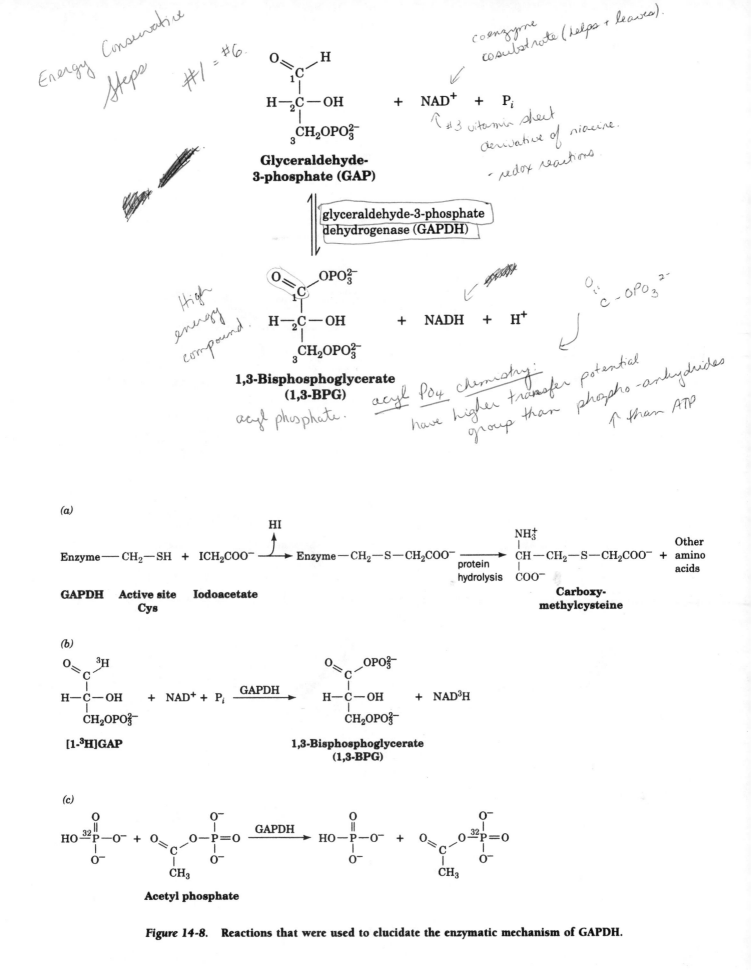

coenzyme cosubstrate (helps + leaves).

O=C—H
|
H—C—OH   + NAD⁺ + Pᵢ
|
CH₂OPO₃²⁻

Glyceraldehyde-3-phosphate (GAP)

#3 vitamin sheet derivative of niacine. - redox reactions

glyceraldehyde-3-phosphate dehydrogenase (GAPDH)

High energy compound.

O=C—OPO₃²⁻
|
H—C—OH   + NADH + H⁺
|
CH₂OPO₃²⁻

1,3-Bisphosphoglycerate (1,3-BPG)

acyl phosphate.   acyl PO₄ chemistry: have higher transfer potential group than phospho-anhydrides ↑ than ATP

(a)

Enzyme—CH₂—SH + ICH₂COO⁻ —(HI)→ Enzyme—CH₂—S—CH₂COO⁻ —(protein hydrolysis)→ CH(NH₃⁺)(COO⁻)—CH₂—S—CH₂COO⁻ + Other amino acids

GAPDH   Active site   Iodoacetate
Cys

Carboxy-methylcysteine

(b)

O=C—³H
|
H—C—OH   + NAD⁺ + Pᵢ —(GAPDH)→
|
CH₂OPO₃²⁻

[1-³H]GAP

O=C—OPO₃²⁻
|
H—C—OH   + NAD³H
|
CH₂OPO₃²⁻

1,3-Bisphosphoglycerate (1,3-BPG)

(c)

HO—³²P(=O)(O⁻)—O⁻ + O=C(CH₃)—O—P(=O)(O⁻)—... —(GAPDH)→ HO—P(=O)(O⁻)—O⁻ + O=C(CH₃)—O—³²P(=O)(O⁻)

Acetyl phosphate

Figure 14-8.   Reactions that were used to elucidate the enzymatic mechanism of GAPDH.

149

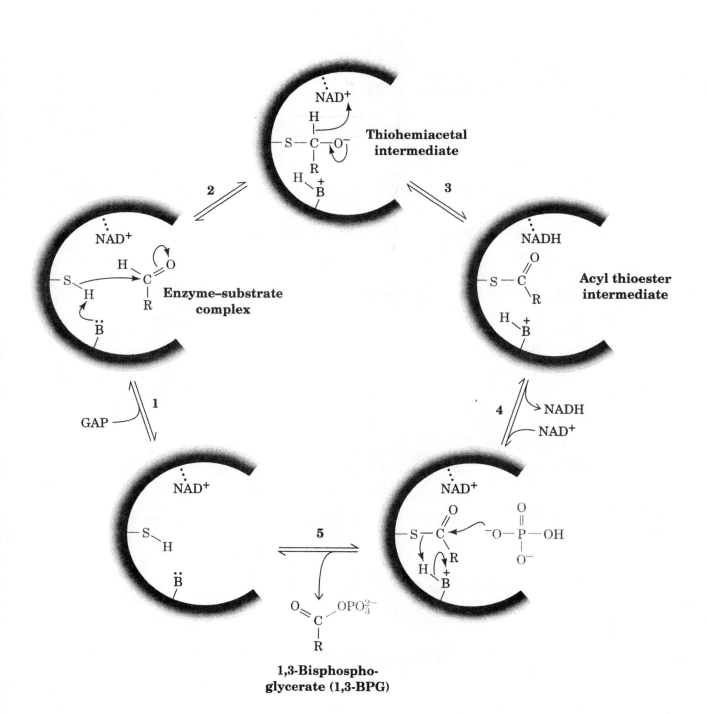

Figure 14-9. The enzymatic mechanism of GAPDH.

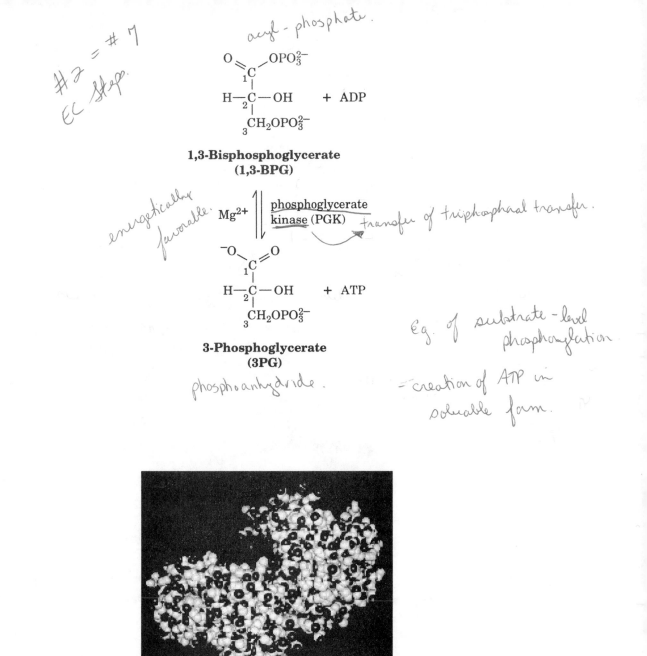

$$\underset{3}{\underset{|}{\overset{\overset{\displaystyle O}{\parallel}}{\underset{|}{C}}}\underset{\displaystyle CH_2OPO_3^{2-}}{\overset{\displaystyle OPO_3^{2-}}{}}}$$

H—C—OH   + ADP

1,3-Bisphosphoglycerate
(1,3-BPG)

Mg²⁺ ⇅ | phosphoglycerate
kinase (PGK)

⁻O—C=O

H—C—OH   + ATP

CH₂OPO₃²⁻

3-Phosphoglycerate
(3PG)

Handwritten annotations:
#2 = #7
EC step.
acyl-phosphate.
energetically favorable.
transfer of triphosphoral transfer.
Eg. of substrate-level phosphorylation.
—creation of ATP in soluble form.
phosphoanhydride.

Figure 14-10.   A space-filling model of yeast phosphoglycerate kinase.

$$GAP + P_i + NAD^+ \longrightarrow 1{,}3\text{-BPG} + NADH \qquad \Delta G^{\circ\prime} = +6.7\,\text{kJ}\cdot\text{mol}^{-1}$$

$$\underline{1{,}3\text{-BPG} + ADP \longrightarrow 3PG + ATP \qquad\qquad \Delta G^{\circ\prime} = -18.8\,\text{kJ}\cdot\text{mol}^{-1}}$$

$$GAP + P_i + NAD^+ + ADP \longrightarrow 3PG + NADH + ATP$$

$$\Delta G^{\circ\prime} = -12.1\,\text{kJ}\cdot\text{mol}^{-1}$$

151

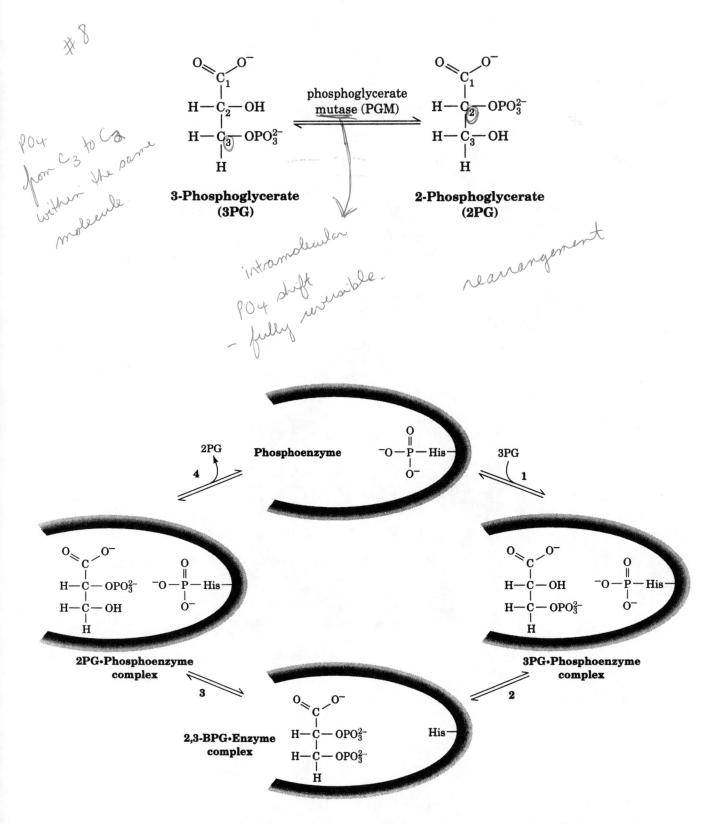

Figure 14-12. **A proposed reaction mechanism for phosphoglycerate mutase.**

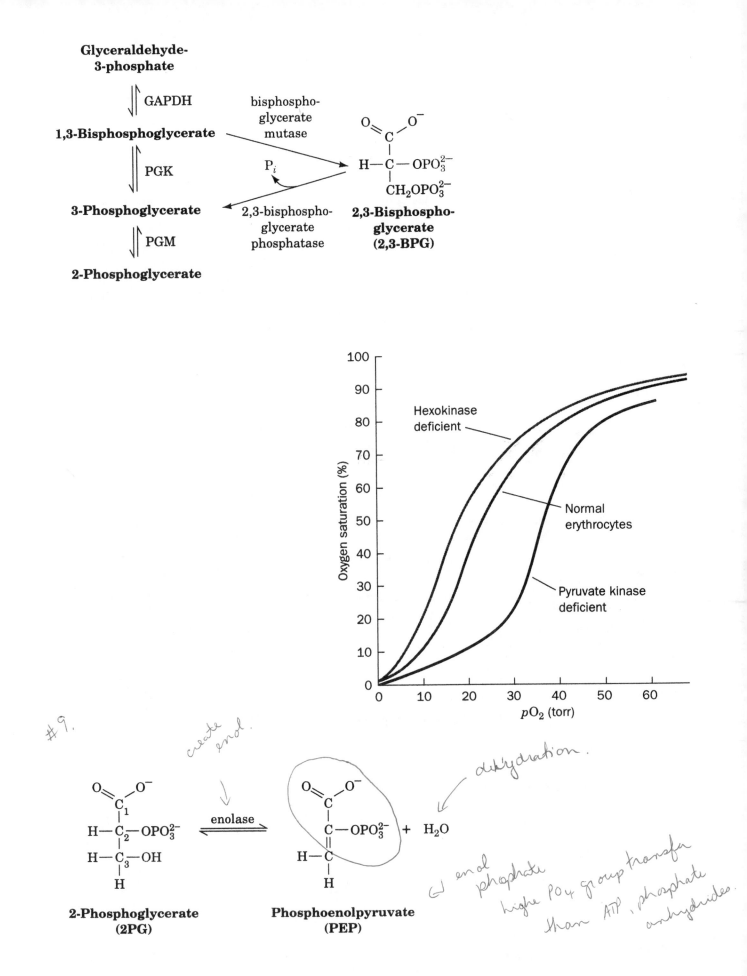

Glyceraldehyde-3-phosphate

$\Updownarrow$ GAPDH

1,3-Bisphosphoglycerate $\longrightarrow$ bisphospho-glycerate mutase

$\Updownarrow$ PGK

3-Phosphoglycerate $\longleftarrow$ 2,3-bisphospho-glycerate phosphatase $\quad P_i$

$\Updownarrow$ PGM

2-Phosphoglycerate

$$O=\overset{\displaystyle O^-}{\underset{\displaystyle |}{C}}$$
$$H-\overset{|}{\underset{|}{C}}-OPO_3^{2-}$$
$$CH_2OPO_3^{2-}$$

2,3-Bisphospho-glycerate (2,3-BPG)

Hexokinase deficient

Normal erythrocytes

Pyruvate kinase deficient

Oxygen saturation (%) vs pO_2 (torr)

#9.

create enol.

$$O=\overset{\displaystyle O^-}{\underset{\displaystyle |}{\underset{1}{C}}}$$
$$H-\overset{|}{\underset{2}{C}}-OPO_3^{2-}$$
$$H-\overset{|}{\underset{3}{C}}-OH$$
$$\overset{|}{H}$$

2-Phosphoglycerate (2PG)

$\overset{\text{enolase}}{\rightleftharpoons}$

$$O=\overset{\displaystyle O^-}{\underset{\displaystyle |}{C}}$$
$$C-OPO_3^{2-}$$
$$H-\overset{\|}{C}$$
$$\overset{|}{H}$$
$+ \; H_2O$

Phosphoenolpyruvate (PEP)

dehydration.

← enol phosphate
higher PO4 group transfer
than ATP, phosphate
anhydrides.

153

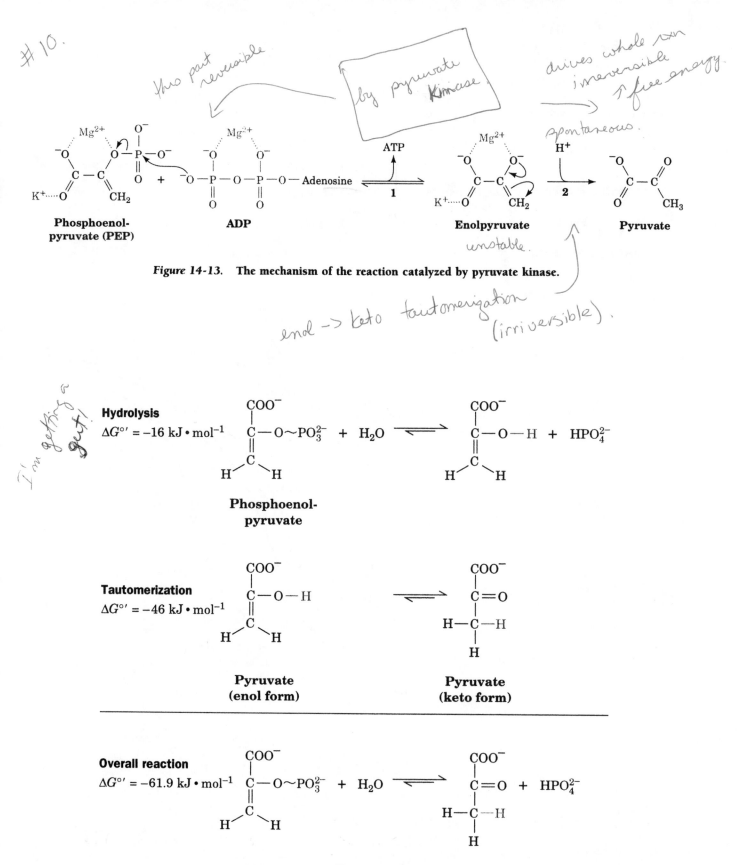

#10.

this part reversible

by pyruvate Kinase

drives whole rxn irreversible ↑free energy.

spontaneous.

Figure 14-13. The mechanism of the reaction catalyzed by pyruvate kinase.

enol → keto tautomerization (irreversible).

unstable.

I'm getting a get!

Hydrolysis
$\Delta G^{\circ\prime} = -16 \text{ kJ} \cdot \text{mol}^{-1}$

Phosphoenol-
pyruvate

Tautomerization
$\Delta G^{\circ\prime} = -46 \text{ kJ} \cdot \text{mol}^{-1}$

Pyruvate
(enol form)

Pyruvate
(keto form)

Overall reaction
$\Delta G^{\circ\prime} = -61.9 \text{ kJ} \cdot \text{mol}^{-1}$

Figure 14-14. The hydrolysis of PEP.

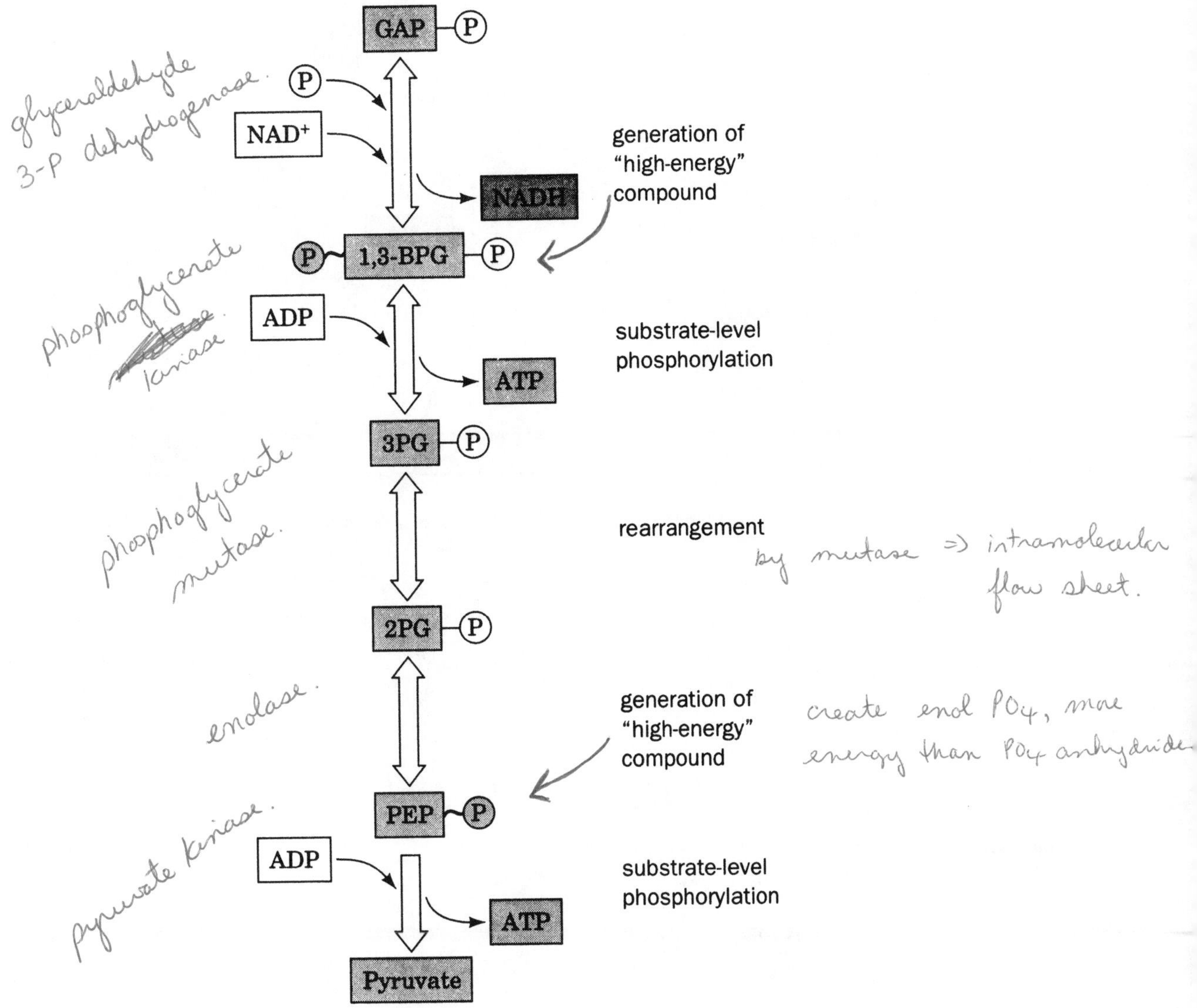

GAP
P
P
NAD+
NADH
generation of
"high-energy"
compound
P
1,3-BPG
P
ADP
ATP
substrate-level
phosphorylation
3PG
P
rearrangement
2PG
P
generation of
"high-energy"
compound
PEP
P
ADP
ATP
substrate-level
phosphorylation
Pyruvate
glyceraldehyde
3-P dehydrogenase.
phosphoglycerate
kinase
phosphoglycerate
mutase.
by mutase => intramolecular
flow sheet.
enolase.
create enol PO4, more
energy than PO4 anhydride
pyruvate kinase.

Figure 14-15. **Schematic view of the second stage of glycolysis.**

$$\text{Glucose} + 2\ \text{NAD}^+ + 2\ \text{ADP} + 2\ \text{P}_i \longrightarrow 2\ \text{pyruvate} + 2\ \text{NADH} + 2\ \text{ATP} + 2\ \text{H}_2\text{O} + 4\ \text{H}^+$$

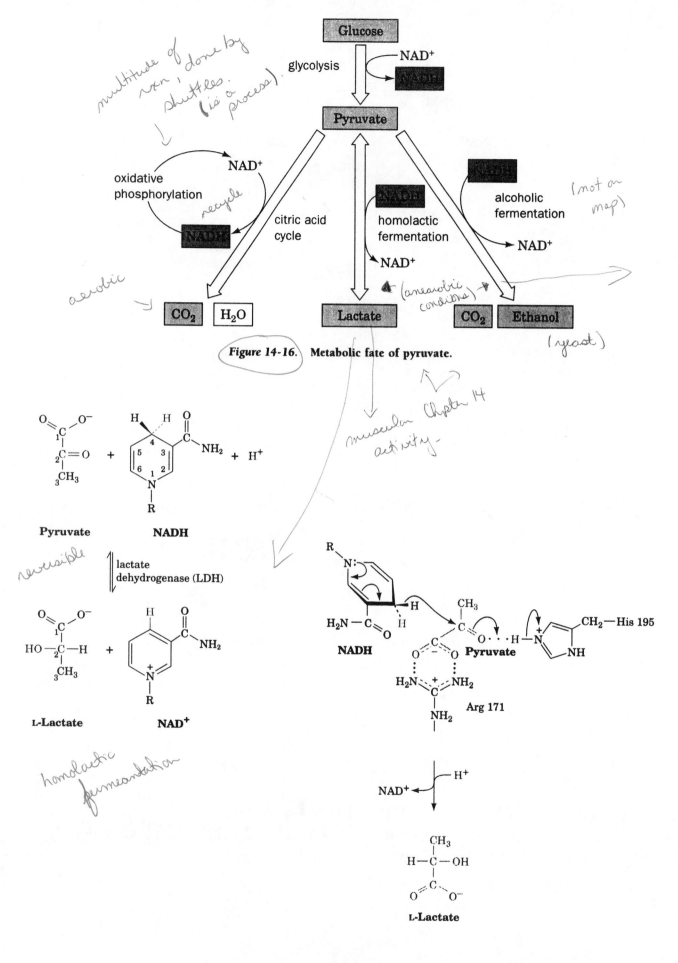

Figure 14-16. Metabolic fate of pyruvate.

(handwritten annotations)
multitude of rxn, done by shuttles. (is a process).

oxidative phosphorylation — recycle

aerobic

(aneaerobic conditions)

(not on map)

(yeast)

muscular activity — Chpter 14

reversible

homolactic fermeantation

Pyruvate   **NADH**

lactate dehydrogenase (LDH)

L-Lactate   **NAD⁺**

NADH   **Pyruvate**   His 195

Arg 171

L-Lactate

156

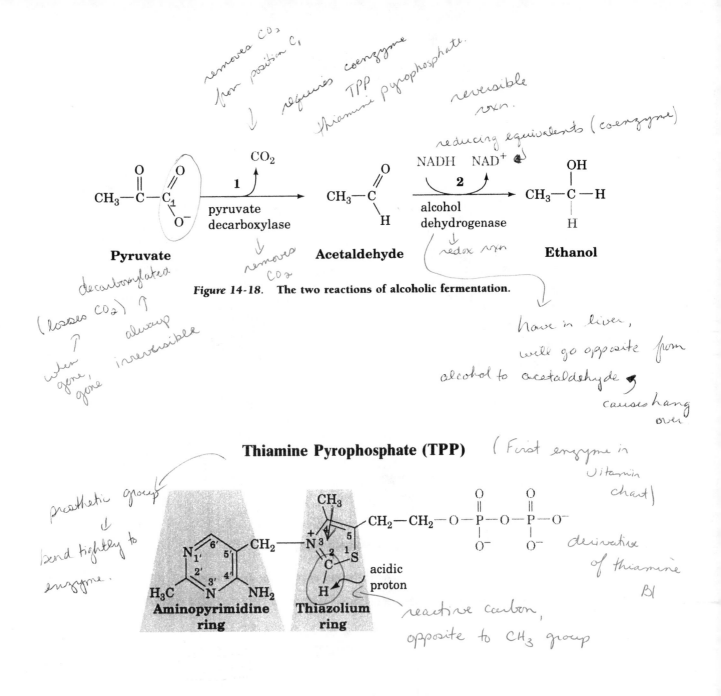

removes CO_2 from position C_1

requires coenzyme TPP thiamine pyrophosphate.

reversible rxn.

reducing equivalents (coenzyme)

Figure 14-18. The two reactions of alcoholic fermentation.

decarboxylated (losses CO_2) when gene, gone, always irreversible

removes CO_2

have in liver, will go opposite from alcohol to acetaldehyde causes hang over

prosthetic group bond tightly to enzyme.

Thiamine Pyrophosphate (TPP)

(First enzyme in vitamin chart)

derivative of thiamine B1

reactive carbon, opposite to CH_3 group

Aminopyrimidine ring

Thiazolium ring

acidic proton

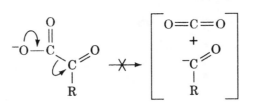

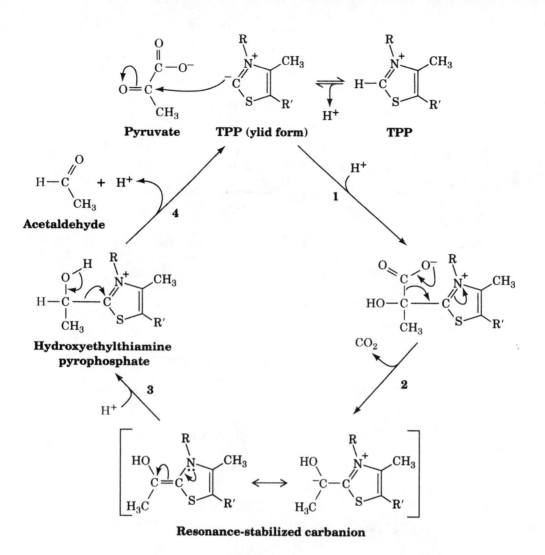

Figure 14-20. The reaction mechanism of pyruvate decarboxylase.

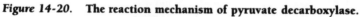

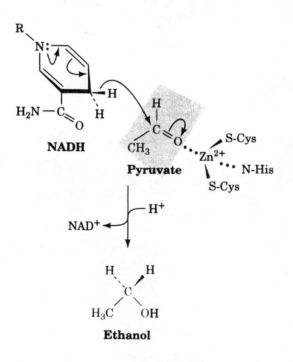

Table 14-1. ΔG°′ and ΔG for the Reactions of Glycolysis in Heart Muscle[a]

Reaction mol^{-1})	Enzyme	ΔG°′ (kJ · mol^{-1})	ΔG (kJ ·
1	Hexokinase	−20.9	−27.2
2	PGI	+2.2	−1.4
3	PFK	−17.2	−25.9
4	Aldolase	+22.8	−5.9
5	TIM	+7.9	+4.4
6 + 7	GAPDH + PGK	−16.7	−1.1
8	PGM	+4.7	−0.6
9	Enolase	−3.2	−2.4
10	PK	−23.0	−13.9

[a]Calculated from data in Newsholme, E.A. and Start, C., *Regulation in Metabolism*, p. 97, Wiley (1973).

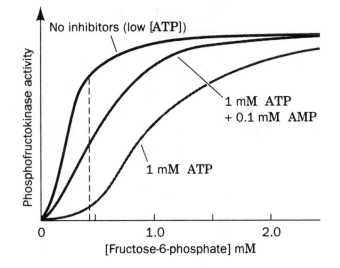

Figure 14-22. PFK activity versus F6P concentration.

Figure 14-21. The X-ray structure of PFK from *E. coli*.

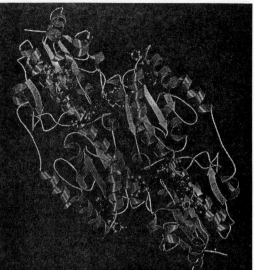

Figure 14-21. The X-ray structure of PFK from *E. coli*.

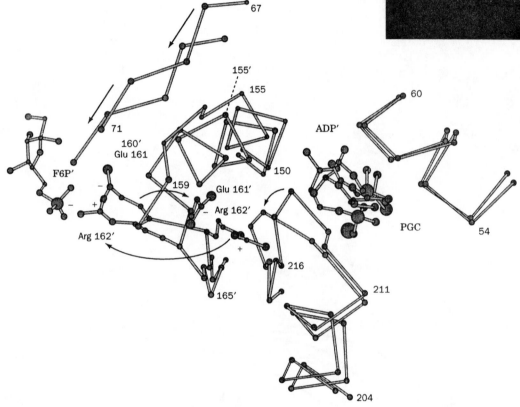

Figure 14-23. Allosteric changes in PFK from *Bacillus stearothermophilus*.

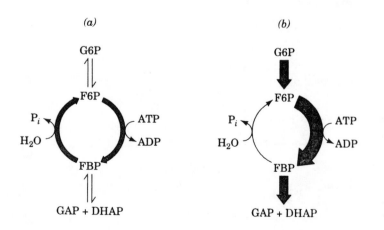

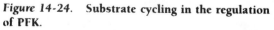

Figure 14-24. Substrate cycling in the regulation of PFK.

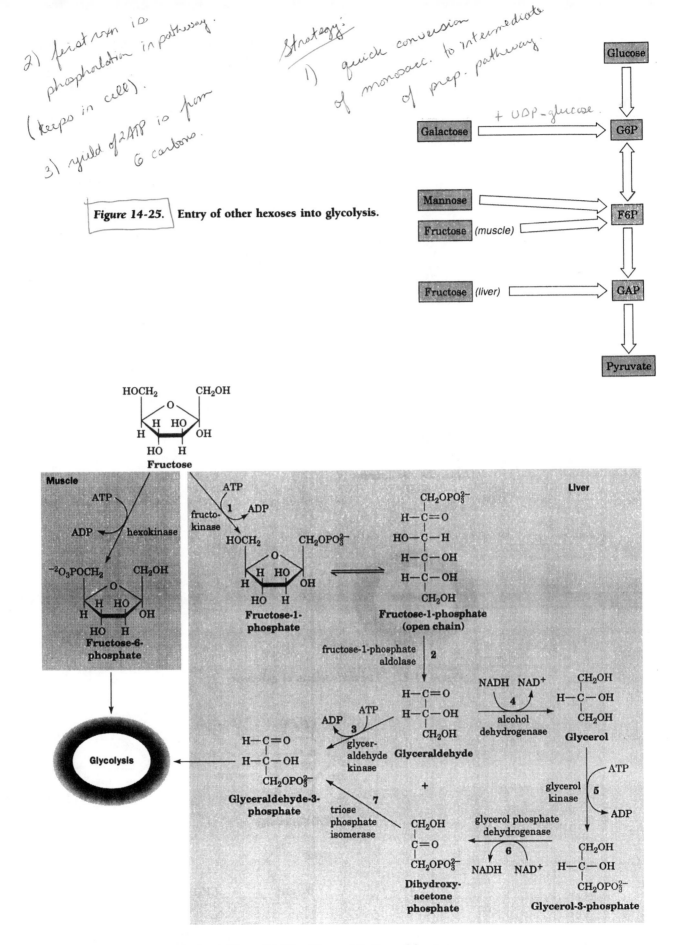

2) first rxn is phosphorelation in pathway. (keeps in cell).

3) yield of 2 ATP is from 6 carbons.

Strategy:

1) quick conversion of monosacc. to intermediate of prep. pathway.

Figure 14-25. Entry of other hexoses into glycolysis.

Figure 14-26. The metabolism of fructose.

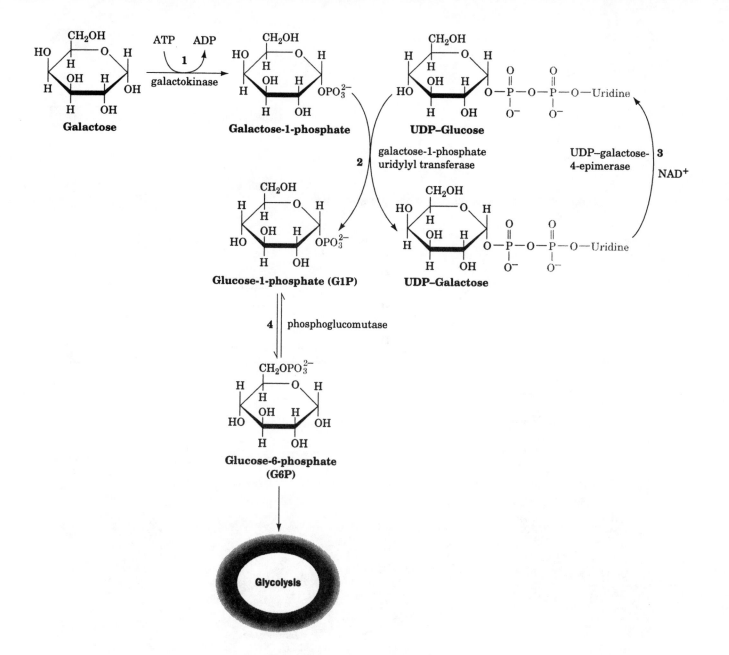

Figure 14-27. The metabolism of galactose.

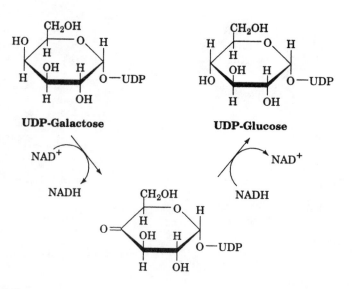

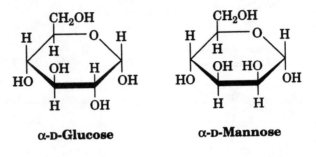

α-D-Glucose   α-D-Mannose

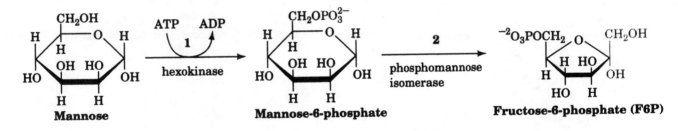

Mannose   Mannose-6-phosphate   Fructose-6-phosphate (F6P)

Figure 14-28. **The metabolism of mannose.**

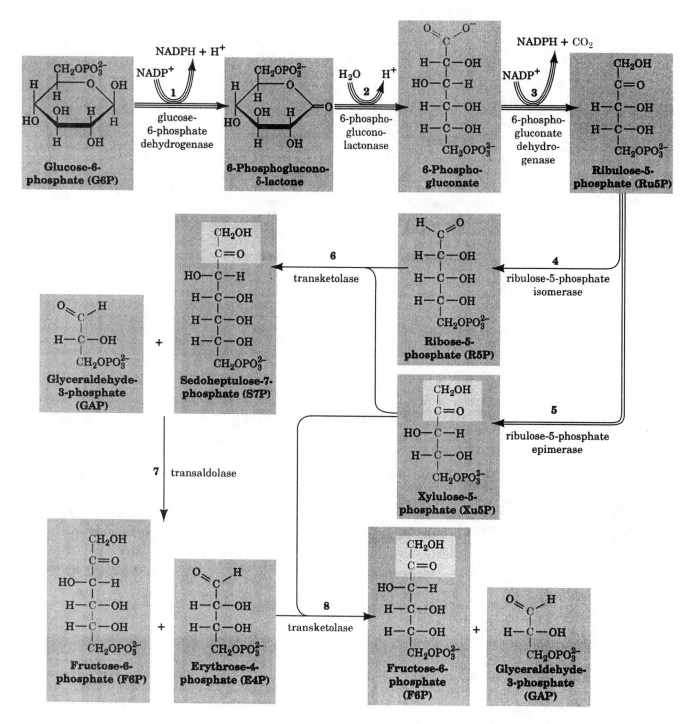

Figure 14-29. *Key to Metabolism.* The pentose phosphate pathway.

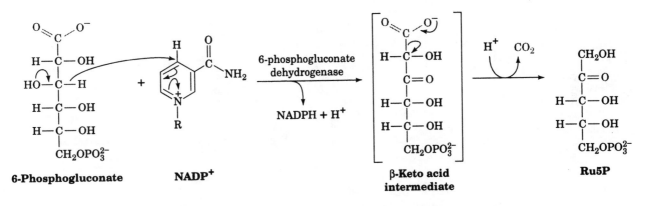

Figure 14-30. The 6-phosphogluconate dehydrogenase reaction.

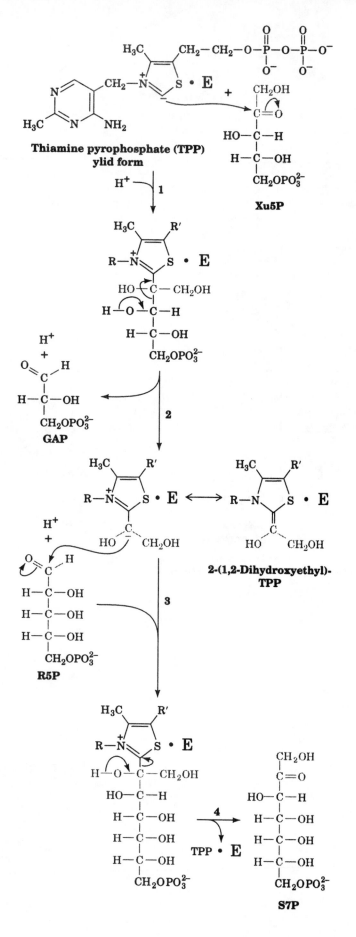

Figure 14-31. Mechanism of transketolase.

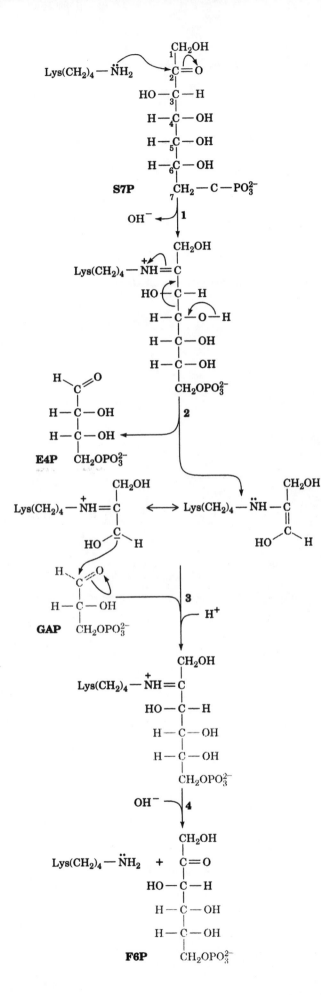

Figure 14-32. Mechanism of transaldolase.

$$\text{(6)} \quad C_5 + C_5 \rightleftharpoons C_7 + C_3$$

$$\text{(7)} \quad C_7 + C_3 \rightleftharpoons C_6 + C_4$$

$$\text{(8)} \quad \underline{C_5 + C_4 \rightleftharpoons C_6 + C_3}$$

$$\text{(Sum)} \quad 3\,C_5 \rightleftharpoons 2\,C_6 + C_3$$

Figure 14-33. **Summary of carbon skeleton rearrangements in the pentose phosphate pathway.**

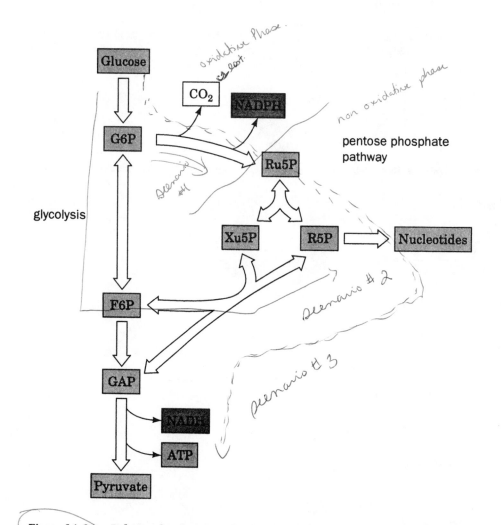

Figure 14-34. **Relationship between glycolysis and the pentose phosphate pathway.**

GLYCOGEN METABOLISM AND GLUCONEOGENESIS

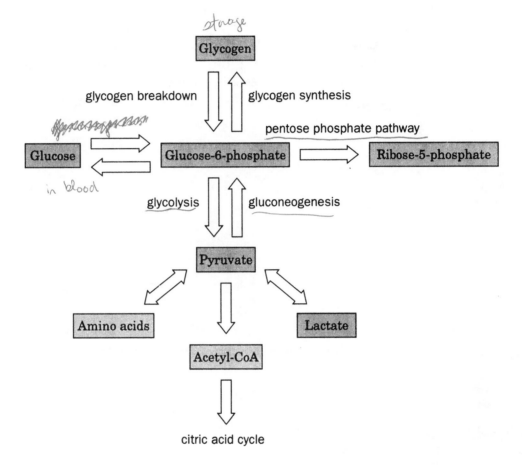

Figure 15-1. **Overview of glucose metabolism.**

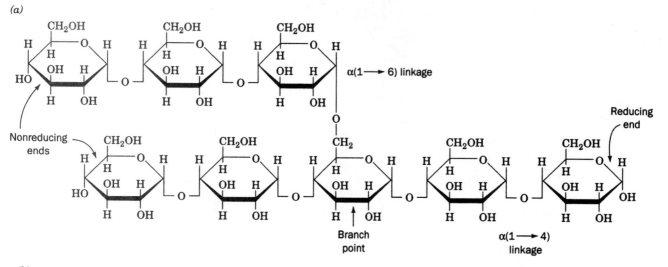

α(1 → 6) linkage

Nonreducing ends

Reducing end

Branch point

α(1 → 4) linkage

(a)

(b)

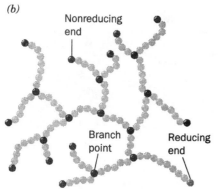

Nonreducing end

Branch point

Reducing end

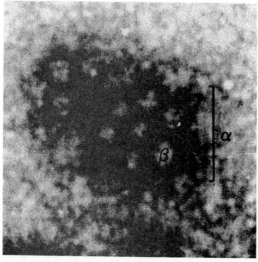

(c)

Figure 15-2. **The structure of glycogen.**

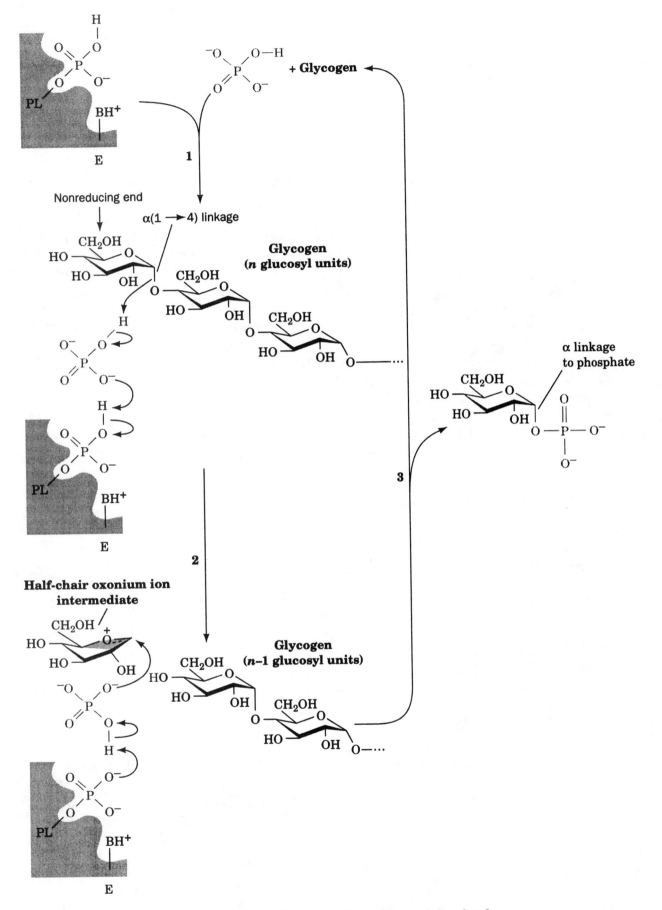

Figure 15-4. The reaction mechanism of glycogen phosphorylase.

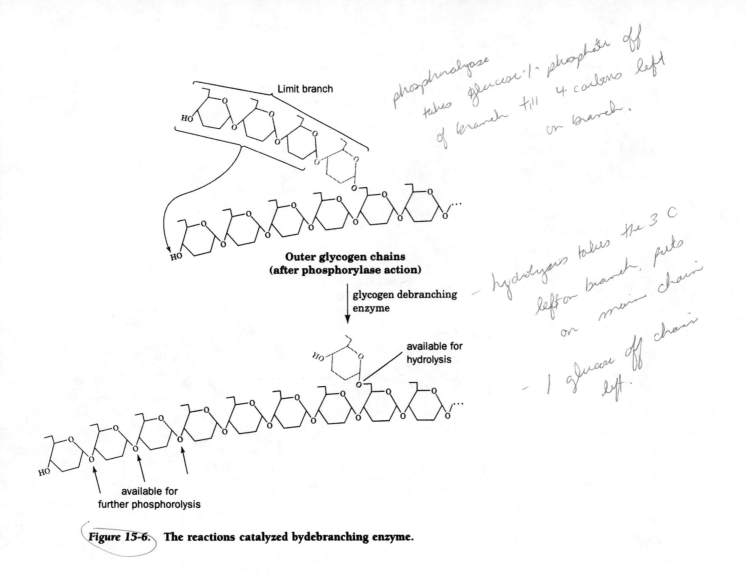

Limit branch

HO

**Outer glycogen chains
(after phosphorylase action)**

glycogen debranching
enzyme

available for
hydrolysis

HO

available for
further phosphorolysis

handwritten margin notes:
phosphorylase takes glucose-1-phosphate off of branch till 4 carbons left on branch.

– hydrolysis takes the 3 C left on branch, puts on main chain

– 1 glucose off chain left.

Figure 15-6. The reactions catalyzed by debranching enzyme.

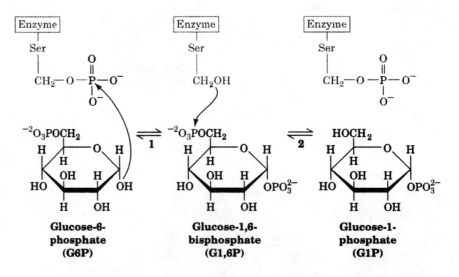

Figure 15-7. The mechanism of phosphoglucomutase.

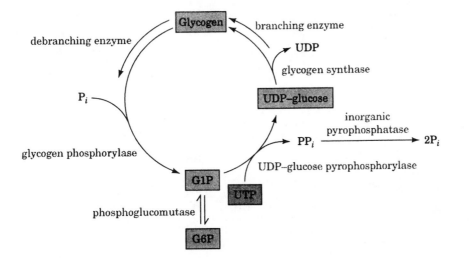

Figure 15-8. Opposing pathways of glycogen synthesis and degradation.

$$\underline{\Delta G^{\circ\prime} \ (\text{kJ} \cdot \text{mol}^{-1})}$$

	$\Delta G^{\circ\prime}$
G1P + UTP $\rightleftharpoons$ UDPG + PP$_i$	~ 0
H$_2$O + PP$_i$ $\longrightarrow$ 2 P$_i$	-33.5
Overall G1P + UTP + H$_2$O $\longrightarrow$ UDPG + 2 P$_i$	-33.5

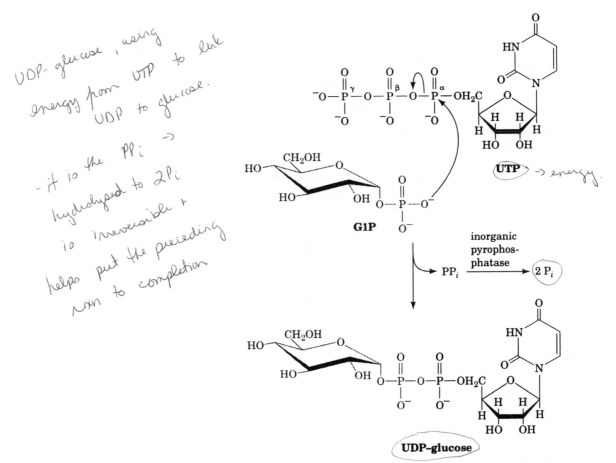

UDP-glucose, using energy from UTP to link UDP to glucose.

- It is the PP$_i$ → hydrolysed to 2P$_i$ is irreversible + helps put the preceding rxn to completion

Figure 15-9. The reaction catalyzed by UDP–glucose pyrophosphorylase.

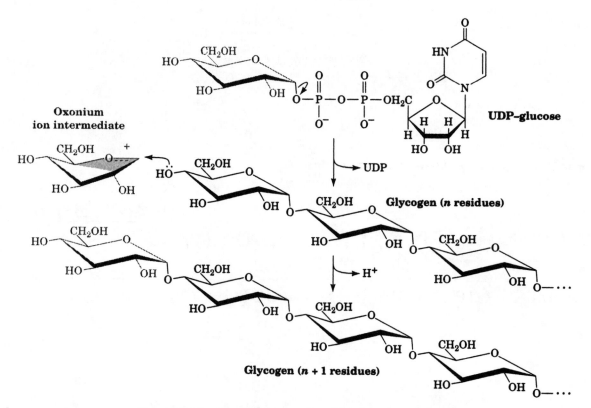

Oxonium ion intermediate

UDP-glucose

Glycogen (*n* residues)

Glycogen (*n* + 1 residues)

Figure 15-10. **The reaction catalyzed by glycogen synthase.**

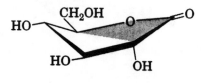

1,5-Gluconolactone

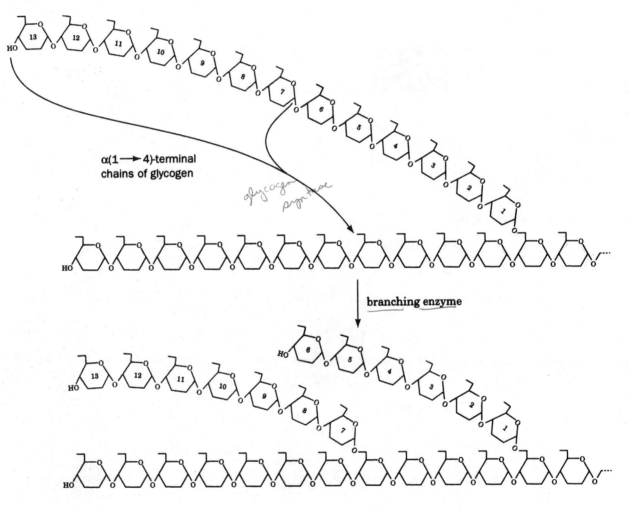

α(1 ⟶ 4)-terminal
chains of glycogen

glycogen synthase

branching enzyme

Figure 15-11. **The branching of glycogen.**

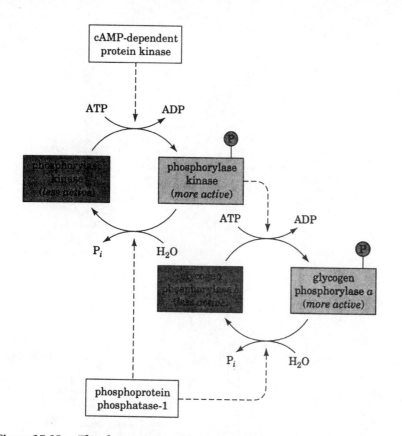

Figure 15-12. The glycogen phosphorylase interconvertible enzyme system.

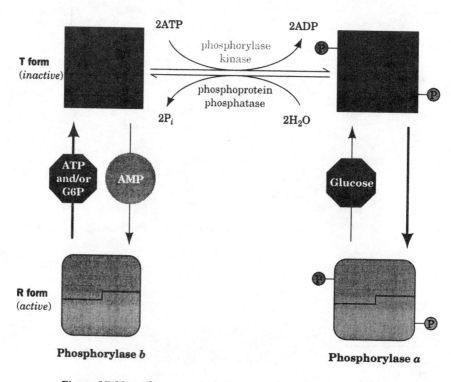

Figure 15-13. The control of glycogen phosphorylase activity.

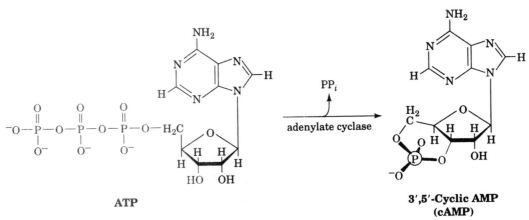

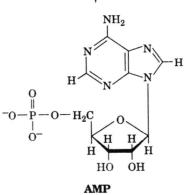

ATP

**3′,5′-Cyclic AMP
(cAMP)**

AMP

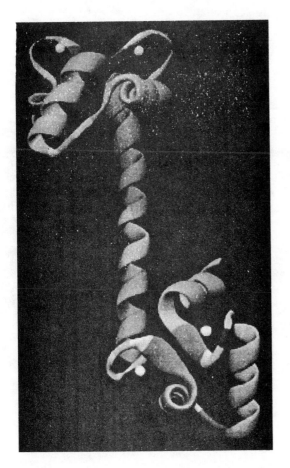

Figure 15-16. The X-ray structure of rat testis calmodulin.

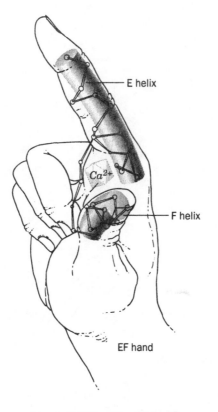

E helix

Ca²⁺

F helix

EF hand

Figure 15-17. **The EF hand.**

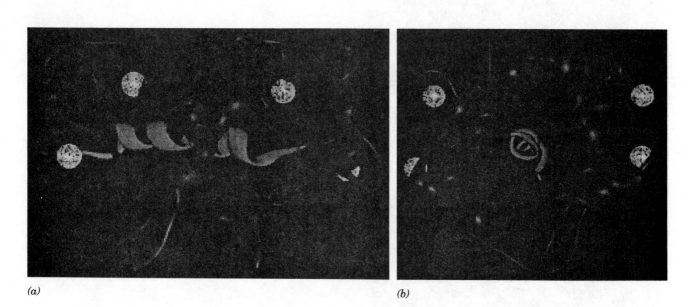

(a)

(b)

Figure 15-18. **The NMR structure of calmodulin in complex with a 26-residue target polypeptide.**

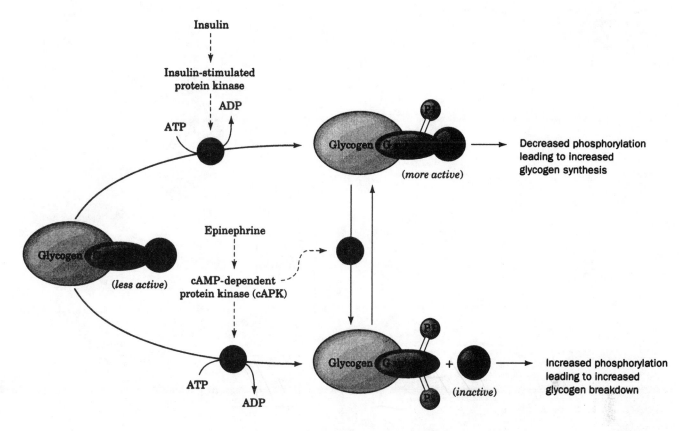

Figure 15-19. Regulation of phosphoprotein phosphatase-1 in muscle.

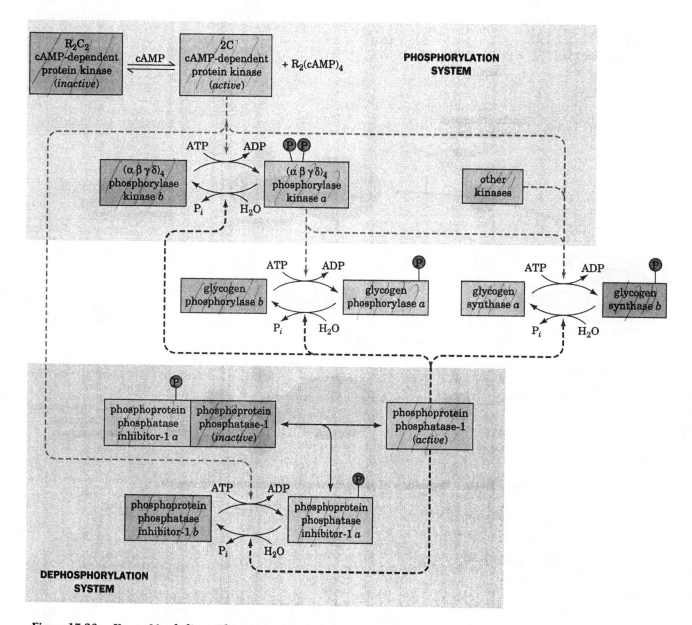

Figure 15-20. *Key to Metabolism.* The major phosphorylation and dephosphorylation systems that regulate glycogen metabolism in muscle.

/// inactive

//// active

$\overset{+}{H_3N}$—His—Ser —Glu—Gly —Thr—Phe—Thr —Ser —Asp—Tyr— 10

Ser—Lys —Tyr—Leu—Asp—Ser —Arg —Arg — Ala— Gln— 20

Asp—Phe—Val—Gln —Trp —Leu—Met—Asn—Thr—COO⁻ 29

Glucagon

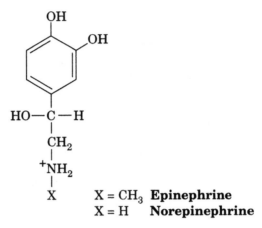

X = CH₃ **Epinephrine**
X = H   **Norepinephrine**

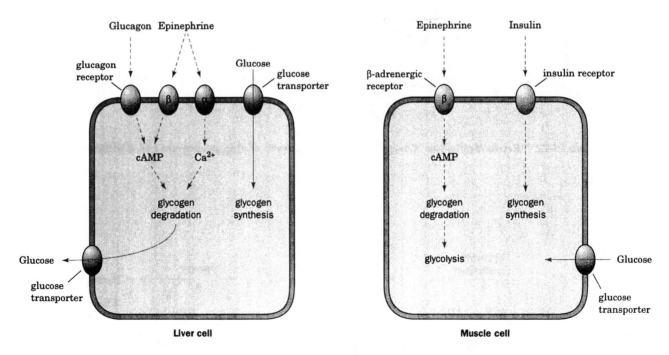

Figure 15-21. **Hormonal control of glycogen metabolism.**

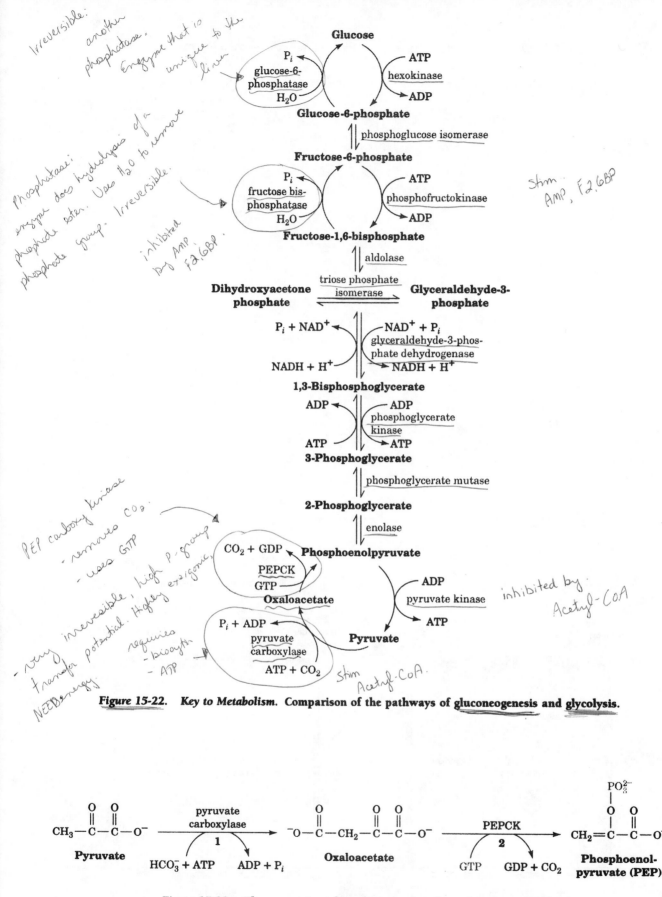

Irreversible: another phosphatase. Enzyme that is unique to the liver

Phosphatase: enzyme does hydrolysis of a phosphate ester. Uses H_2O to remove phosphate group. Irreversible

inhibited by AMP, F2,6BP

Glucose

P_i ← glucose-6-phosphatase / H_2O ATP — hexokinase → ADP

Glucose-6-phosphate

phosphoglucose isomerase

Fructose-6-phosphate

P_i ← fructose bis-phosphatase / H_2O ATP — phosphofructokinase → ADP

Stim: AMP, F2,6BP

Fructose-1,6-bisphosphate

aldolase

triose phosphate isomerase

Dihydroxyacetone phosphate   **Glyceraldehyde-3-phosphate**

P_i + NAD$^+$ ← / NADH + H$^+$ NAD$^+$ + P_i — glyceraldehyde-3-phosphate dehydrogenase → NADH + H$^+$

1,3-Bisphosphoglycerate

ADP ← / ATP   ADP — phosphoglycerate kinase → ATP

3-Phosphoglycerate

phosphoglycerate mutase

2-Phosphoglycerate

enolase

PEP carboxy kinase
- removes CO_2
- uses GTP

- very irreversible, high P group transfer potential. Highly exergonic, requires NEEDs energy
- biotin
- ATP

CO_2 + GDP ← / PEPCK / GTP —   **Phosphoenolpyruvate**

Oxaloacetate   ADP — pyruvate kinase → ATP

P_i + ADP ← pyruvate carboxylase / ATP + CO_2 →   **Pyruvate**

inhibited by: Acetyl-CoA

Stim Acetyl-CoA

Figure 15-22. *Key to Metabolism.* **Comparison of the pathways of gluconeogenesis and glycolysis.**

$$CH_3-\overset{O}{\underset{}{C}}-\overset{O}{\underset{}{C}}-O^-$$

Pyruvate

pyruvate carboxylase **1**

HCO_3^- + ATP   ADP + P_i

$$^-O-\overset{O}{\underset{}{C}}-CH_2-\overset{O}{\underset{}{C}}-\overset{O}{\underset{}{C}}-O^-$$

Oxaloacetate

PEPCK **2**

GTP   GDP + CO_2

$$CH_2=\overset{PO_3^{2-}}{\underset{}{C}}-\overset{O}{\underset{}{C}}-O^-$$

Phosphoenol-pyruvate (PEP)

Figure 15-23. **The conversion of pyruvate to phosphoenolpyruvate (PEP).**

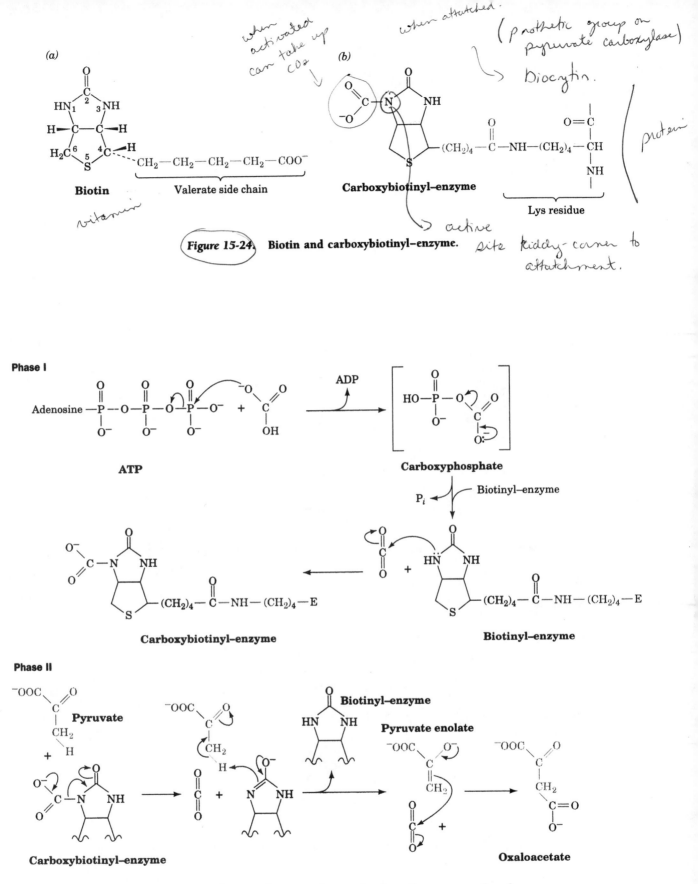

(prothetic group on pyruvate carboxylase)

Figure 15-24. Biotin and carboxybiotinyl–enzyme.

Figure 15-25. The two-phase reaction mechanism of pyruvate carboxylase.

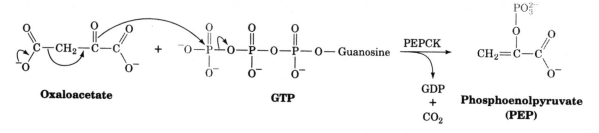

Figure 15-26. The PEPCK mechanism.

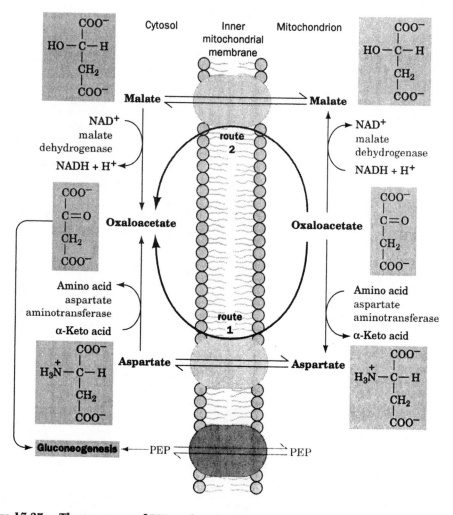

Figure 15-27. The transport of PEP and oxaloacetate from the mitochondrion to the cytosol.

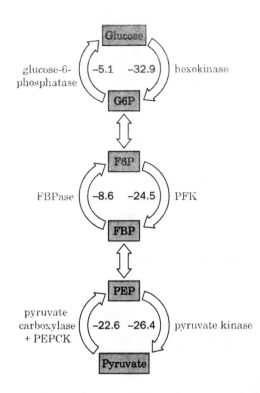

Figure 15-28. Substrate cycles in glucose metabolism.

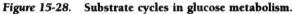

β-D-Fructose-2,6-bisphosphate
(F2,6P)

if molecule not around.
will not stimulate
PFK1,
↓
then gluconeogenesis

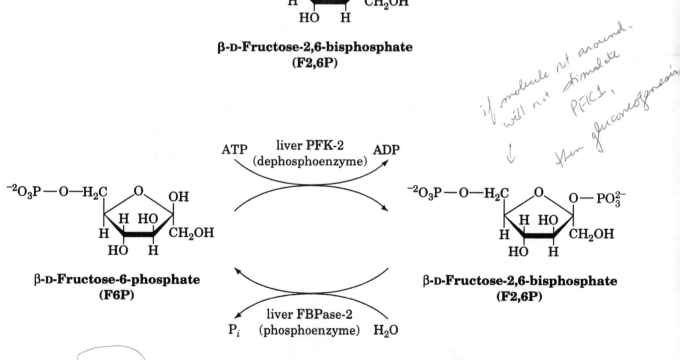

β-D-Fructose-6-phosphate
(F6P)

β-D-Fructose-2,6-bisphosphate
(F2,6P)

Figure 15-29. The formation and degradation of β-D-fructose-2,6-bisphosphate (F2,6P).

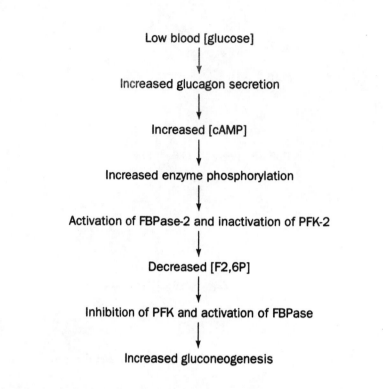

Low blood [glucose]

↓

Increased glucagon secretion

↓

Increased [cAMP]

↓

Increased enzyme phosphorylation

↓

Activation of FBPase-2 and inactivation of PFK-2

↓

Decreased [F2,6P]

↓

Inhibition of PFK and activation of FBPase

↓

Increased gluconeogenesis

Figure 15-30. Sequence of metabolic events linking low blood [glucose] to gluconeogenesis in liver.

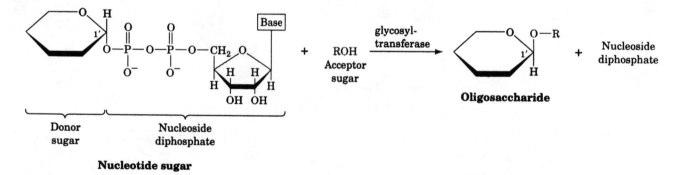

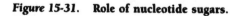

Figure 15-31. Role of nucleotide sugars.

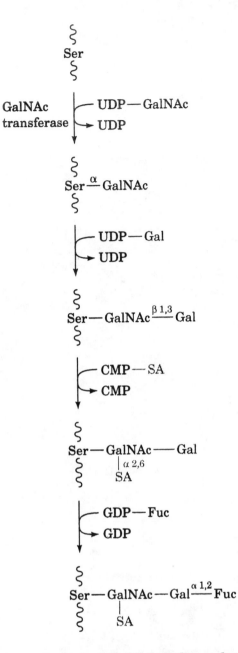

Figure 15-32. Synthesis of an *O*-linked oligosaccharide chain.

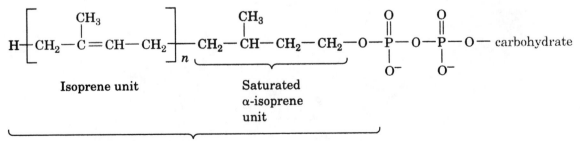

Isoprene unit

Saturated
α-isoprene
unit

Dolichol

Figure 15-33. Dolichol pyrophosphate glycoside.

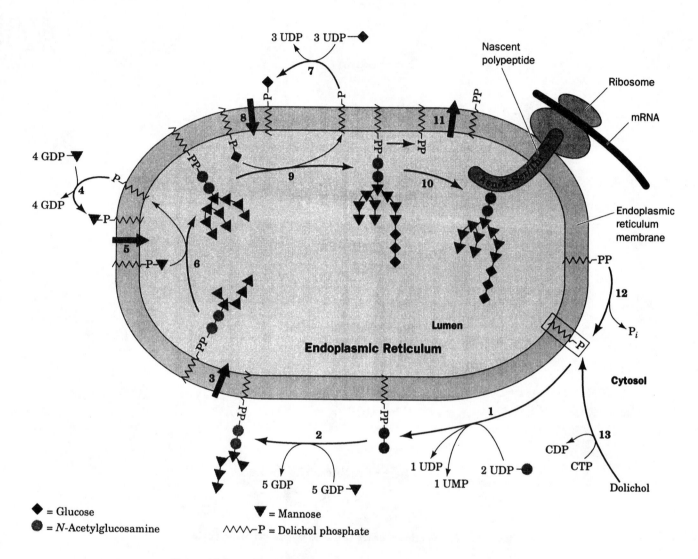

= Glucose

= N-Acetylglucosamine

= Mannose

−P = Dolichol phosphate

Figure 15-34. The pathway of dolichol-PP-oligosaccharide synthesis.

CITRIC ACID CYCLE

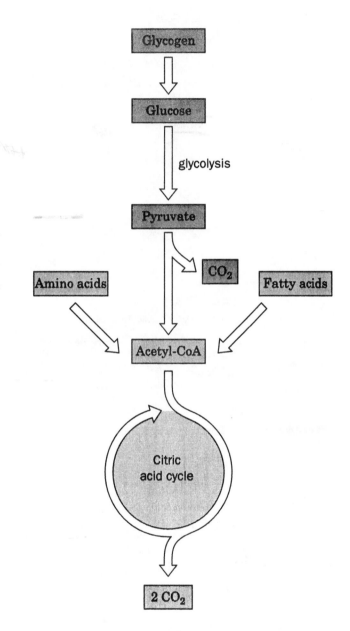

Figure 16-1. Overview of oxidative fuel metabolism.

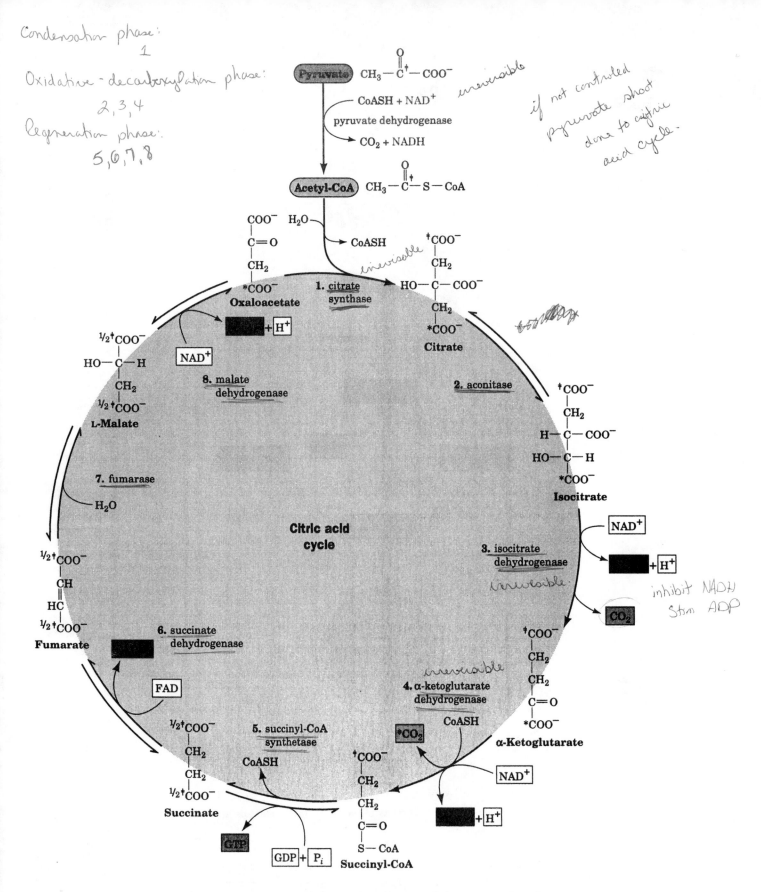

Condensation phase:
1

Oxidative-decarboxylation phase:
2, 3, 4

Regeneration phase:
5, 6, 7, 8

irreversible

if not controlled
pyruvate shoot
done to citric
acid cycle.

inhibit NADH
Stim ADP

Figure 16-2. *Key to Metabolism.* **The reactions of the citric acid cycle.**

190

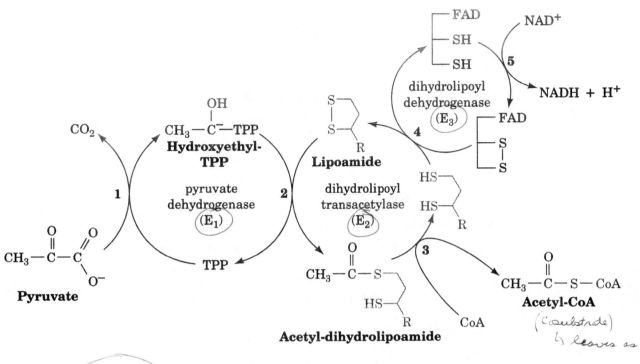

Figure 16-5. The five reactions of the pyruvate dehydrogenase multienzyme complex.

Table 16-1. The Coenzymes and Prosthetic Groups of Pyruvate Dehydrogenase

Cofactor	Location	Function
Thiamine pyrophosphate (TPP)	Bound to E_1	Decarboxylates pyruvate yielding a hydroxyethyl-TPP carbanion
Lipoic acid	Covalently linked to a Lys on E_2 (lipoamide)	Accepts the hydroxyethyl carbanion from TPP as an acetyl group
Coenzyme A (CoA)	Substrate for E_2	Accepts the acetyl group from lipoamide
Flavin adenine dinucleotide (FAD)	Bound to E_3	Reduced by lipoamide
Nicotinamide adenine dinucleotide (NAD^+)	Substrate for E_3	Reduced by $FADH_2$

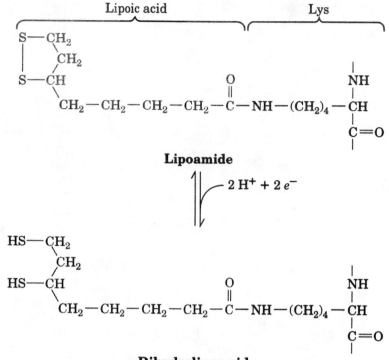

Lipoamide

$2\,H^+ + 2\,e^-$

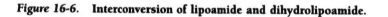

Dihydrolipoamide

Figure 16-6. **Interconversion of lipoamide and dihydrolipoamide.**

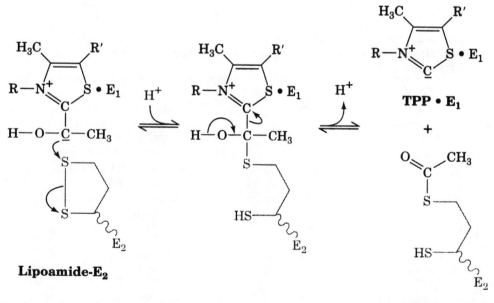

Lipoamide-E₂

TPP • E₁

+

**Acetyl-
dihydrolipoamide-E₂**

Acetyl-CoA

+

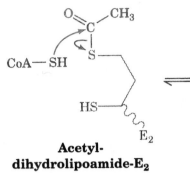

**Acetyl-
dihydrolipoamide-E₂**

Dihydrolipoamide-E₂

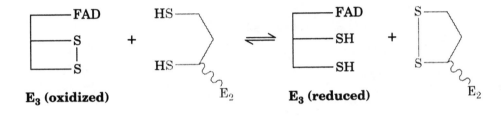

E₃ (oxidized)   **E₃ (reduced)**

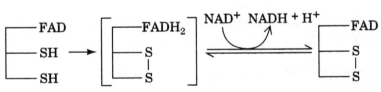

E₃ (oxidized)

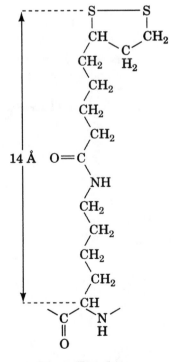

**Lipoyllysyl arm
(fully extended)**

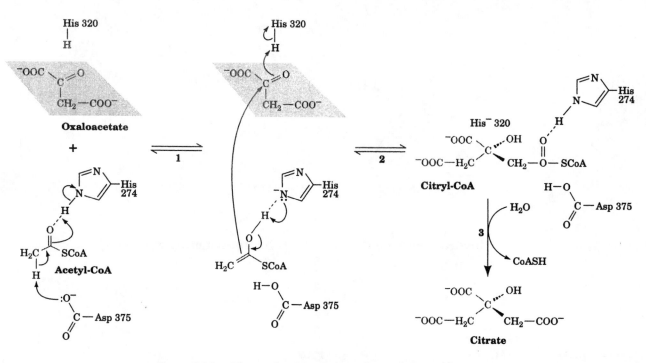

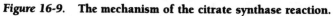

Figure 16-9. The mechanism of the citrate synthase reaction.

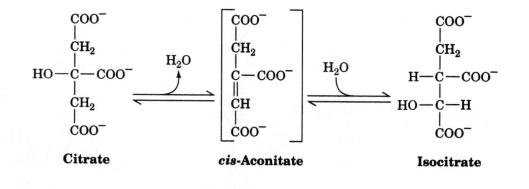

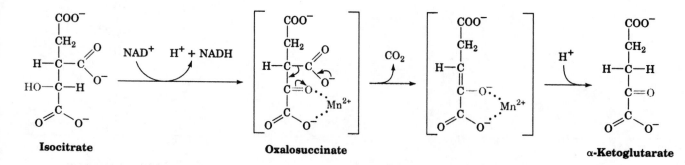

Figure 16-10. The reaction mechanism of isocitrate dehydrogenase.

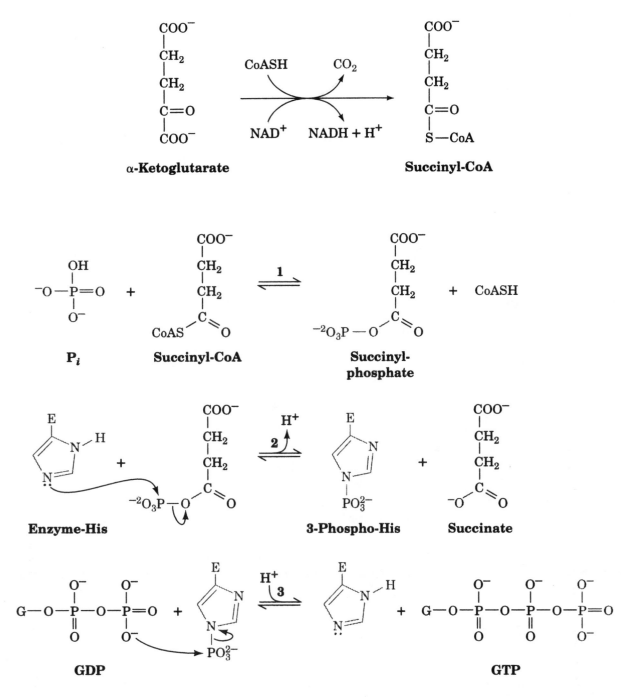

Figure 16-11. The reaction catalyzed by succinyl-CoA synthetase.

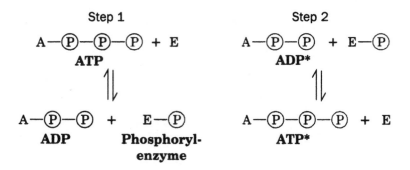

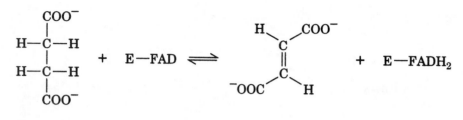

Succinate   **Fumarate**

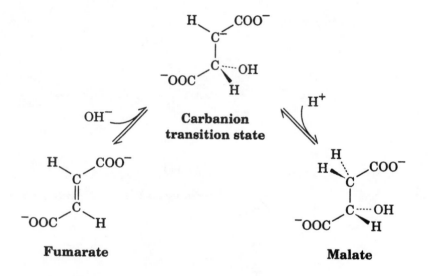

Fumarate   **Carbanion transition state**   **Malate**

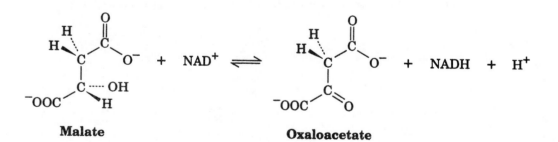

Malate   **Oxaloacetate**

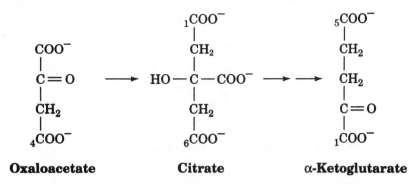

Oxaloacetate **Citrate** **α-Ketoglutarate**

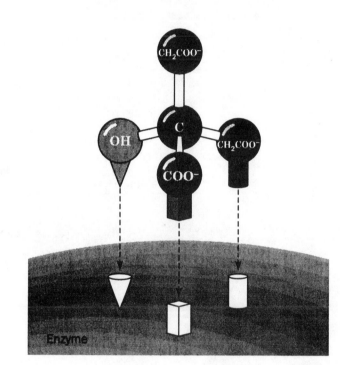

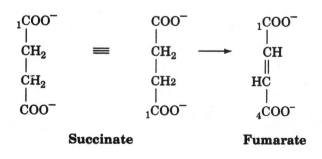

Succinate **Fumarate**

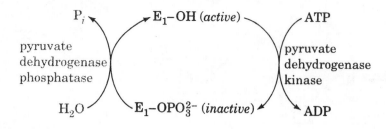

Figure 16-13. Covalent modification of eukaryotic pyruvate dehydrogenase.

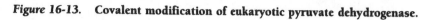

Table 16-2. Standard Free Energy Changes (ΔG°') and Physiological Free Energy Changes (ΔG) of Citric Acid Cycle Reactions

Reaction	Enzyme	ΔG°' (kJ·mol⁻¹)	ΔG (kJ·mol⁻¹)
1	Citrate synthase	−31.5	Negative
2	Aconitase	~5	~0
3	Isocitrate dehydrogenase	−21	Negative
4	α-Ketoglutarate dehydrogenase multienzyme complex	−33	Negative
5	Succinyl-CoA synthetase	−2.1	~0
6	Succinate dehydrogenase	+6	~0
7	Fumarase	−3.4	~0
8	Malate dehydrogenase	+29.7	~0

provide free energy to turn whole cycle.

- are points of control (the 3 rxns).

- occur early in cycle. is evidence of cycle. Put energy anywhere in cycle + will turn cycle.

Figure 16-14. Regulation of the citric acid cycle.

Blue →
    amoplerotic reaction.

Red → various intermediates
    can be used for
    anabolic functions.
= amphibolic pathways
    ↳ does both catabolism
        + make anabolic
            materials

❄ two we look at.

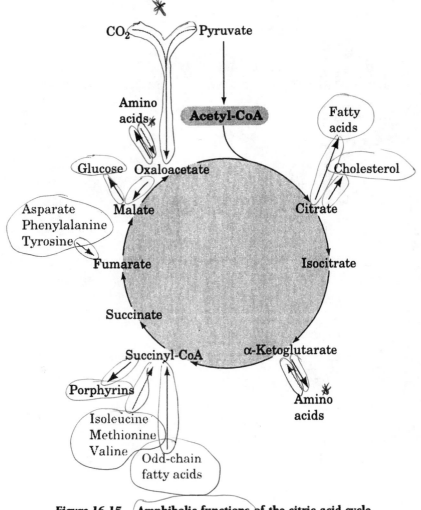

Figure 16-15. Amphibolic functions of the citric acid cycle.

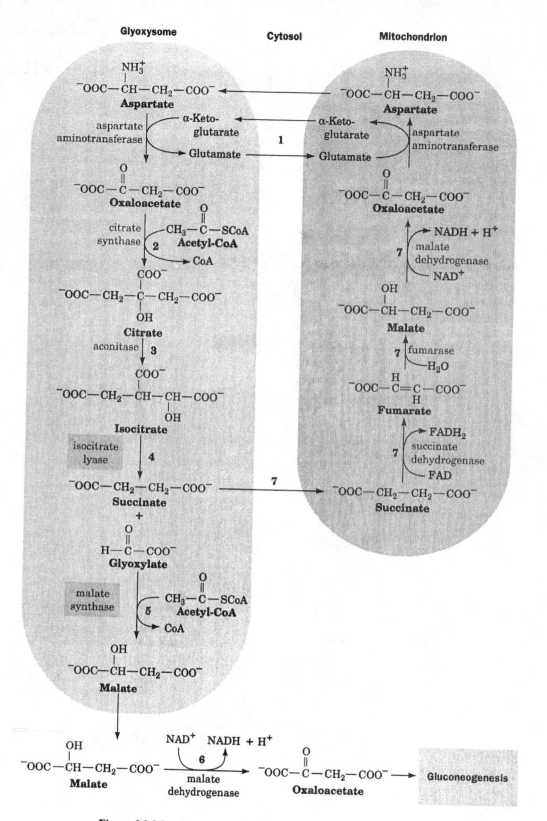

Figure 16-16. *Key to Metabolism.* **The glyoxylate pathway.**

ELECTRON TRANSPORT AND OXIDATIVE PHOSPHORYLATION

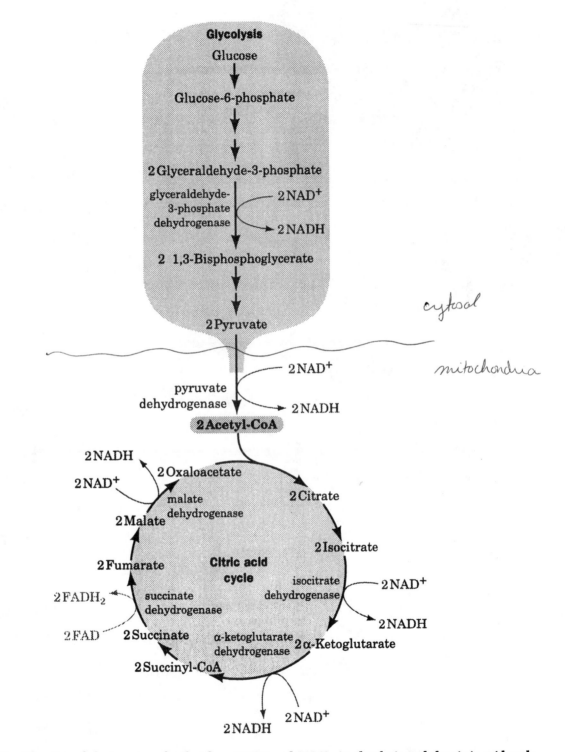

Figure 17-1. The sites of electron transfer that form NADH and FADH$_2$ in glycolysis and the citric acid cycle.

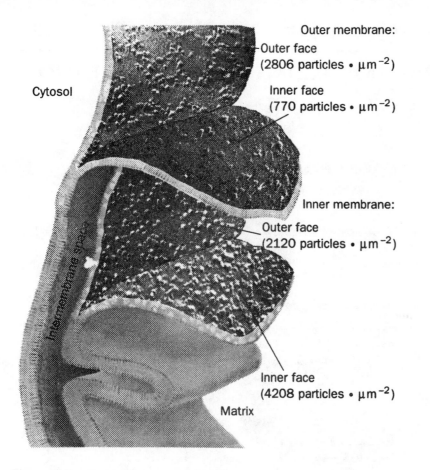

Cytosol

Outer membrane:
Outer face
(2806 particles $\cdot$ μm^{-2})

Inner face
(770 particles $\cdot$ μm^{-2})

Inner membrane:

Outer face
(2120 particles $\cdot$ μm^{-2})

Intermembrane space

Inner face
(4208 particles $\cdot$ μm^{-2})

Matrix

Figure 17-3. Freeze-fracture and freeze-etch electron micrographs of the inner and outer mitochondrial membranes.

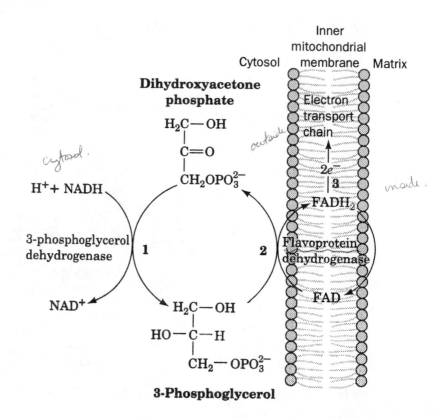

Figure 17-4. The glycerophosphate shuttle.

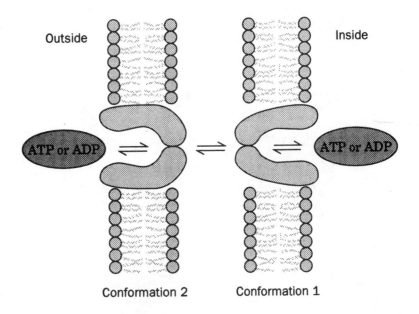

Figure 17-5. Conformational mechanism of the ADP–ATP translocator.

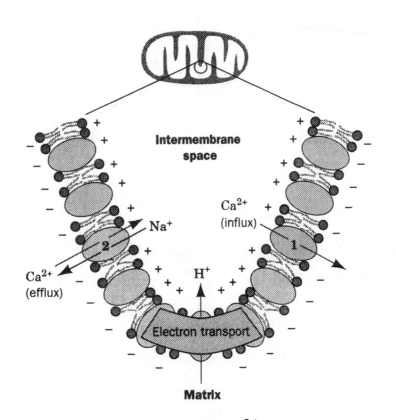

Figure 17-6. The two mitochondrial Ca^{2+} transport systems.

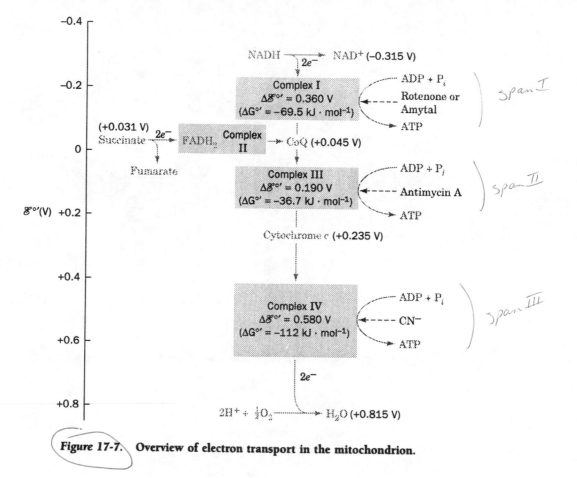

Figure 17-7. Overview of electron transport in the mitochondrion.

$$\tfrac{1}{2} O_2 + NADH + H^+ \rightleftharpoons H_2O + NAD^+$$

$$\Delta\mathscr{E}^{\circ\prime} = 0.815\ V - (-0.315\ V) = 1.130\ V$$

$$\Delta G^{\circ\prime} = -n\mathscr{F}\Delta\mathscr{E}^{\circ\prime}$$

$$NADH + CoQ\ (oxidized) \longrightarrow NAD^+ + CoQ\ (reduced)$$

$$\Delta\mathscr{E}^{\circ\prime} = 0.360\ V \qquad \Delta G^{\circ\prime} = -69.5\ kJ\cdot mol^{-1}$$

$$CoQ\ (reduced) + cytochrome\ c\ (oxidized) \longrightarrow$$
$$CoQ\ (oxidized) + cytochrome\ c\ (reduced)$$

$$\Delta\mathscr{E}^{\circ\prime} = 0.190\ V \qquad \Delta G^{\circ\prime} = -36.7\ kJ\cdot mol^{-1}$$

$$Cytochrome\ c\ (reduced) + \tfrac{1}{2} O_2 \longrightarrow cytochrome\ c\ (oxidized) + H_2O$$

$$\Delta\mathscr{E}^{\circ\prime} = 0.580\ V \qquad \Delta G^{\circ\prime} = -112\ kJ\cdot mol^{-1}$$

Table 17-1. Reduction Potentials of Electron-Transport Chain Components in Resting Mitochondria

Component	$\mathscr{E}\mathscr{E}$ (V)
NADH	−0.315
Complex I (NADH–CoQ reductase; 850 kD, 43 subunits):	
FMN	?
(Fe–S)N-1a	−0.380
(Fe–S)N-1b	−0.250
(Fe–S)N-2	−0.030
(Fe–S)N-3,4	−0.245
(Fe–S)N-5,6	−0.270
Succinate	0.031
Complex II (succinate–CoQ reductase; 127 kD, 5 subunits):	
FAD	−0.040
(Fe–S)S-1	−0.030
(Fe–S)S-2	−0.245
(Fe–S)S-3	0.060
Cytochrome b_{560}	−0.080
Coenzyme Q	0.045
Complex III (CoQ–cytochrome c reductase; 248 kD, 11 subunits):	
Cytochrome b_H (b_{562})	0.030
Cytochrome b_L (b_{566})	−0.030
(Fe–S)	0.280
Cytochrome c_1	0.215
Cytochrome c	0.235
Complex IV (cytochrome c oxidase; ~200 kD, 6–13 subunits):	
Cytochrome a	0.210
Cu_A center	0.245
Cu_B	0.340
Cytochrome a_3	0.385
O_2	0.815

Source: Wilson, D.F., Erecińska, M., and Dutton, P.L., *Annu. Rev. Biophys. Bioeng.* 3, 205 and 208 (1974); *and* Wilson, D.F., *In* Bittar, E.E. (Ed.), *Membrane Structure and Function*, Vol. 1, p. 160, Wiley (1980).

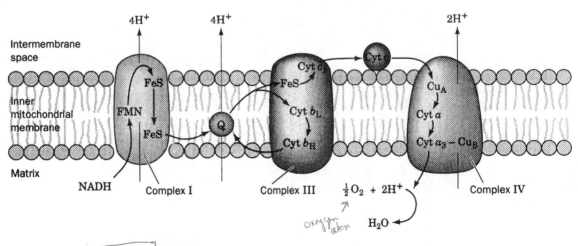

Figure 17-8. *Key to Function.* The mitochondrial electron-transport chain.

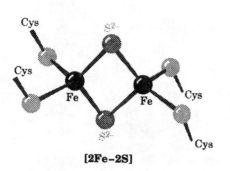

[2Fe–2S]

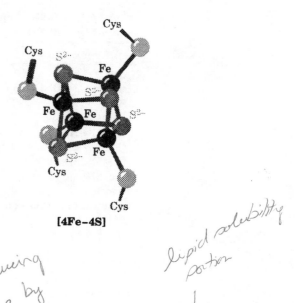

[4Fe–4S]

(a)

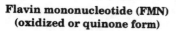

CH₂OPO₃²⁻

Flavin mononucleotide (FMN)
(oxidized or quinone form)

⇅ [H•]

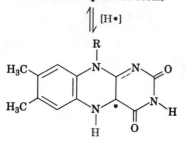

FMNH• (radical or semiquinone form)

⇅ [H•]

FMNH₂ (reduced or hydroquinone form)

(b)

site where reducing agents picked up by quinone

highly mobile

lipid solubility portion

isoprene unit

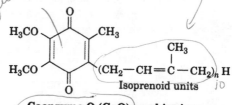

(CH₂—CH=C—CH₂)ₙ H

Isoprenoid units

Coenzyme Q (CoQ) or ubiquinone
(oxidized or quinone form)

⇅ [H•]

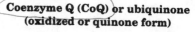

Coenzyme QH• or ubisemiquinone
(radical or semiquinone form)

⇅ [H•]

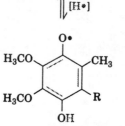

Coenzyme QH₂ or ubiquinol
(reduced or hydroquinone form)

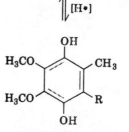

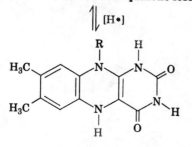

Figure 17-10. **The oxidation states of FMN and coenzyme Q.**

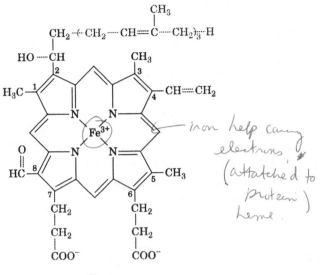

*iron help carry
electrons,
(attached to
protein)
Heme.*

Heme a

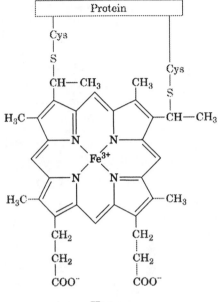

Heme c

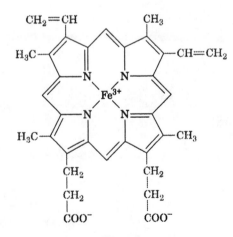

**Heme b
(iron–protoporphyrin IX)**

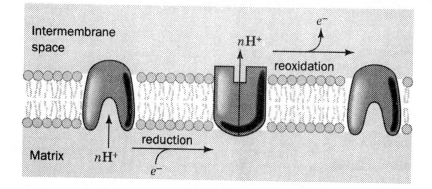

Figure 17-11. A model for electron transport–linked proton pumping.

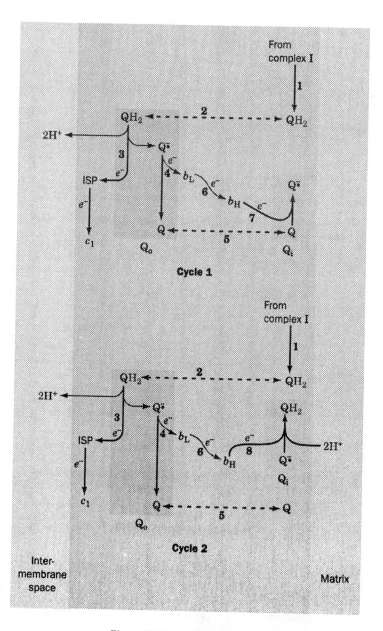

Figure 17-13. The Q cycle.

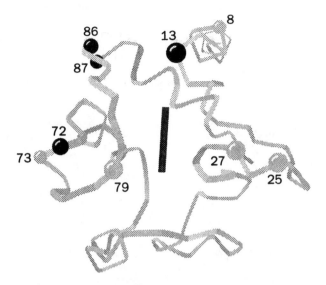

Figure 17-14. Ribbon diagram of cytochrome *c* showing the Lys residues involved in intermolecular complex formation.

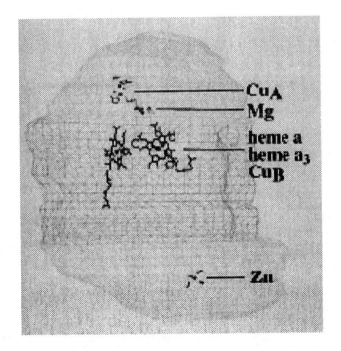

Figure 17-16. Locations of redox centers in beef heart cytochrome *c* oxidase.

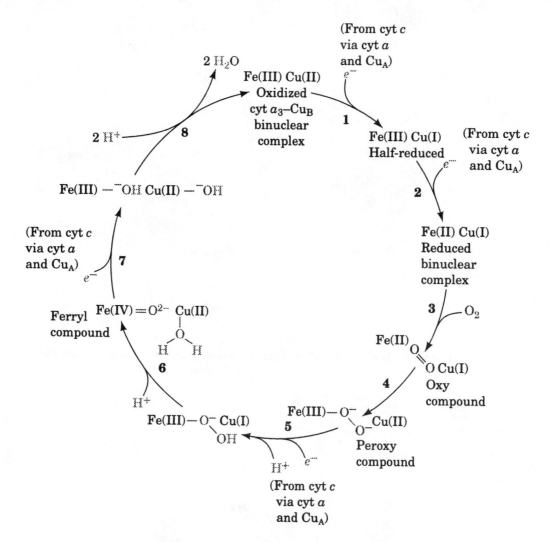

Figure 17-17. The cytochrome *c* oxidase reaction.

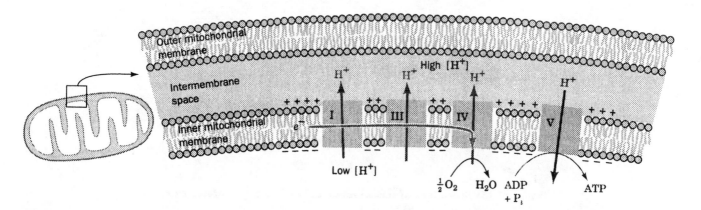

Figure 17-18. The coupling of electron transport and ATP synthesis.

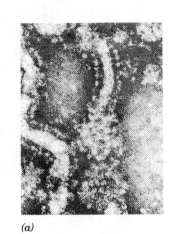

(a)

(b)

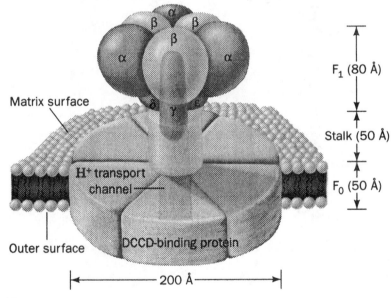

(c)

Figure 17-19. **Structure of ATP synthase.**

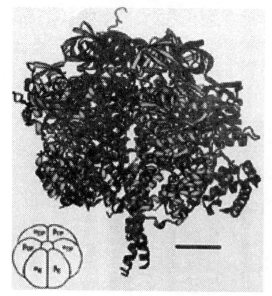

(a)

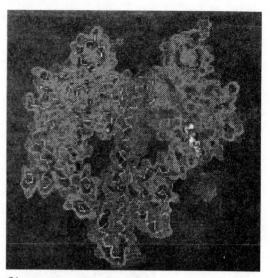

(b)

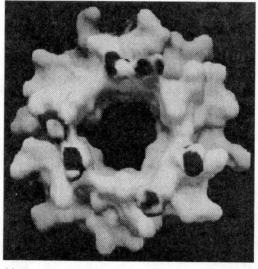

(c)

Figure 17-20. The X-ray structure of F_1-ATPase from bovine heart mitochondria.

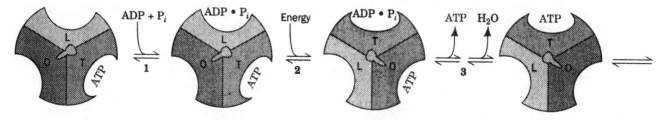

Figure 17-21. *Key to Function.* The binding change mechanism for ATP synthase.

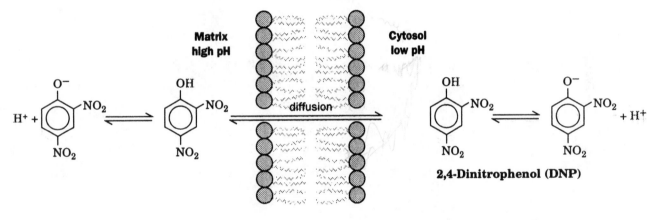

Figure 17-22.   Action of 2,4-dinitrophenol.

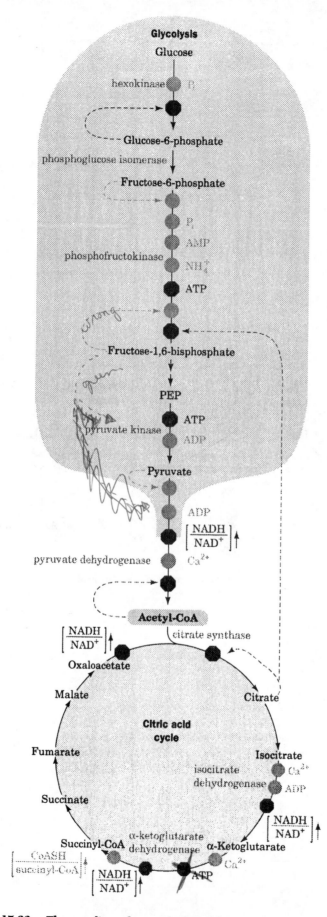

Glycolysis

Glucose

hexokinase   Pᵢ

Glucose-6-phosphate

phosphoglucose isomerase

Fructose-6-phosphate

Pᵢ

AMP

NH₄⁺

phosphofructokinase   ATP

Fructose-1,6-bisphosphate

PEP

ATP

pyruvate kinase   ADP

Pyruvate

ADP

$\left[\dfrac{NADH}{NAD^+}\right]\uparrow$

pyruvate dehydrogenase   Ca²⁺

Acetyl-CoA

citrate synthase

$\left[\dfrac{NADH}{NAD^+}\right]\uparrow$

Oxaloacetate

Malate   Citrate

Fumarate

Citric acid cycle

Succinate   **Isocitrate**

isocitrate dehydrogenase   Ca²⁺   ADP

$\left[\dfrac{NADH}{NAD^+}\right]\uparrow$

Succinyl-CoA   α-ketoglutarate dehydrogenase   α-**Ketoglutarate**   Ca²⁺

$\left[\dfrac{CoASH}{succinyl\text{-}CoA}\right]\uparrow$ $\left[\dfrac{NADH}{NAD^+}\right]\uparrow$ ATP

Figure 17-23. The coordinated control of glycolysis and the citric acid cycle.

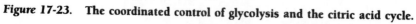

PHOTOSYNTHESIS

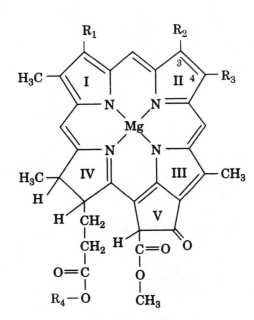

Chlorophyll

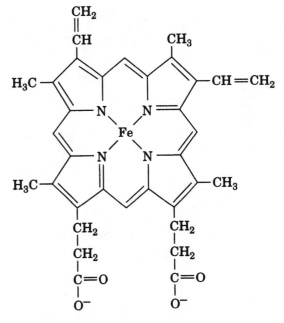

Iron–protoporphyrin IX

	R₁	R₂	R₃	R₄
Chlorophyll *a*	$-CH=CH_2$	$-CH_3$	$-CH_2-CH_3$	P
Chlorophyll *b*	$-CH=CH_2$	$\overset{O}{\overset{\|}{-C}}-H$	$-CH_2-CH_3$	P
Bacteriochlorophyll *a*	$\overset{O}{\overset{\|}{-C}}-CH_3$	$-CH_3$ [a]	$-CH_2-CH_3$ [a]	P or G
Bacteriochlorophyll *b*	$\overset{O}{\overset{\|}{-C}}-CH_3$	$-CH_3$ [a]	$=CH-CH_3$ [a]	P

[a] No double bond between positions C3 and C4.

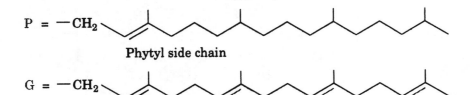

P = —CH₂

Phytyl side chain

G = —CH₂

Geranylgeranyl side chain

Figure 18-2. **Chlorophyll structures.**

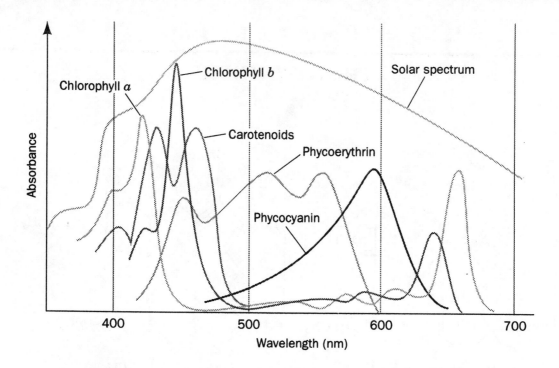

Figure 18-3.   The absorption spectra of various photosynthetic pigments.

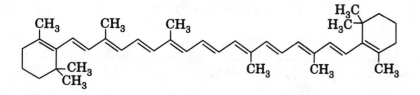

β-Carotene

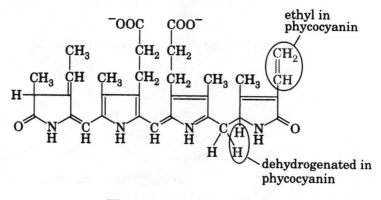

Phycoerythrin

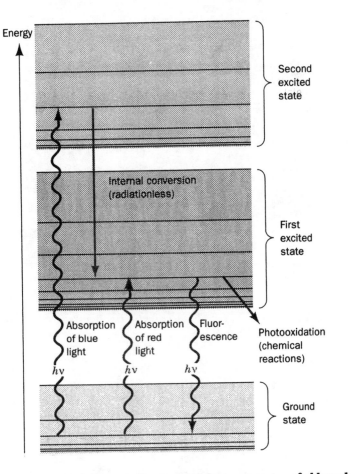

Figure 18-6. An energy diagram indicating the electronic states of chlorophyll and their most important modes of interconversion.

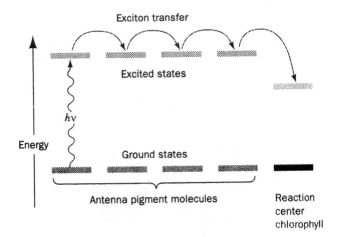

Figure 18-7. Excitation energy trapping by the photosynthetic reaction center.

217

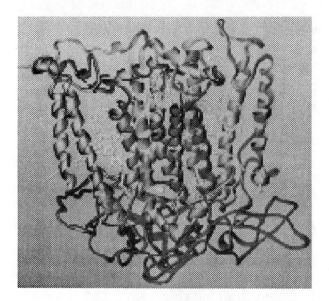

Figure 18-8.   A ribbon diagram of the photosynthetic reaction center from *Rb. sphaeroides*.

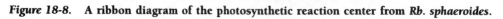

Menaquinone

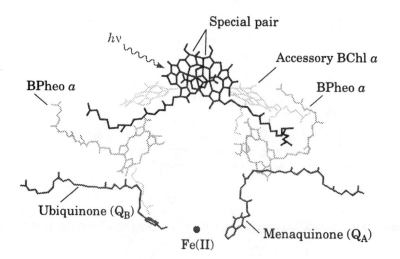

Figure 18-9.   Disposition of prosthetic groups in the photosynthetic reaction center of *Rps. viridis*.

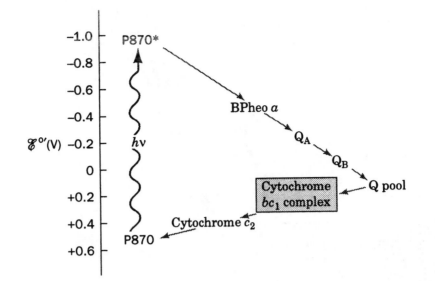

Figure 18-10. The photosynthetic electron-transport system of purple photosynthetic bacteria.

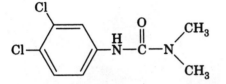

**3-(3,4-Dichlorophenyl)-1,1-dimethylurea
(DCMU)**

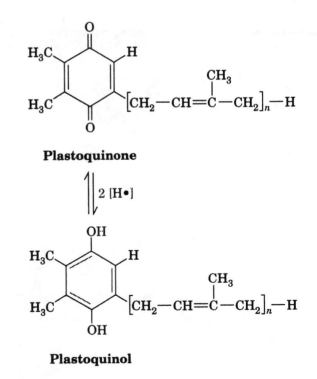

Plastoquinone

2 [H•]

Plastoquinol

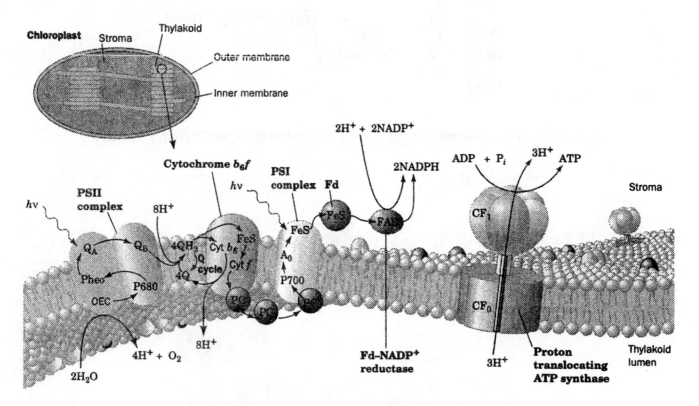

Figure 18-11. *Key to Function.* **A model of the thylakoid membrane.**

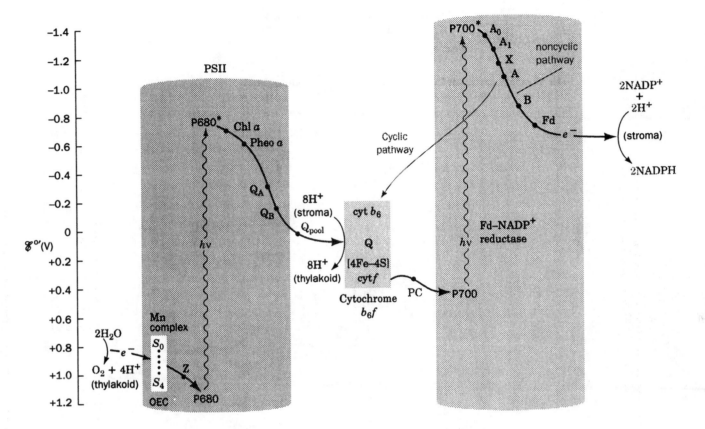

Figure 18-12. The Z-scheme of photosynthesis.

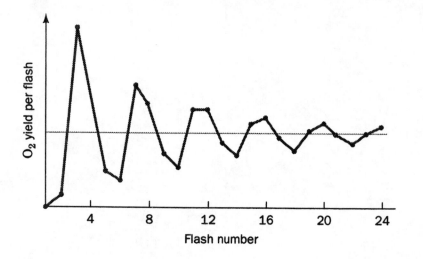

Figure 18-13. The O$_2$ yield per flash in dark-adapted spinach chloroplasts.

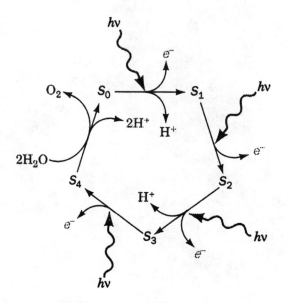

Figure 18-14. The schematic mechanism of O$_2$ generation in chloroplasts.

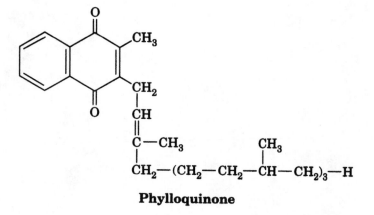

Phylloquinone

222

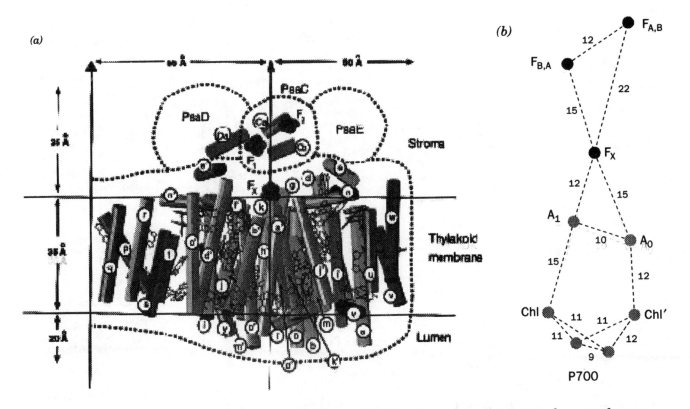

Figure 18-17. The low-resolution structure of a subunit of PSI from the cyanobacterium *Synechococcus elongatus*.

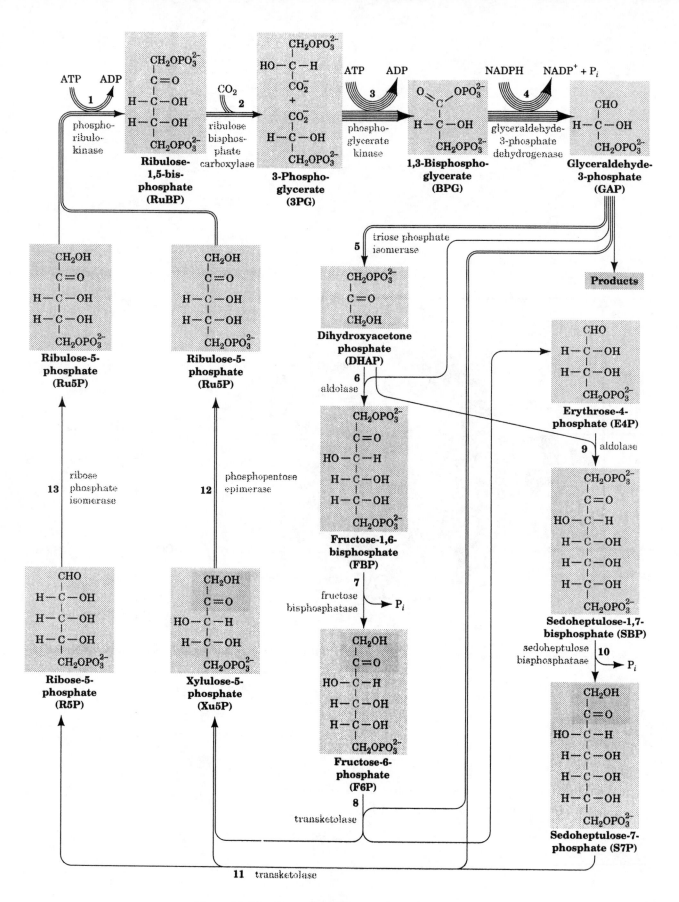

Figure 18-20. Key to Metabolism. The Calvin cycle.

6. $C_3 + C_3 \longrightarrow C_6$
8. $C_3 + C_6 \longrightarrow C_4 + C_5$
9. $C_3 + C_4 \longrightarrow C_7$
11. $C_3 + C_7 \longrightarrow C_5 + C_5$

$$5\,C_3 \longrightarrow 3\,C_5$$

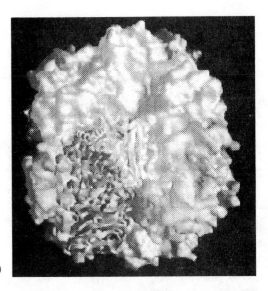

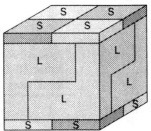

(a)

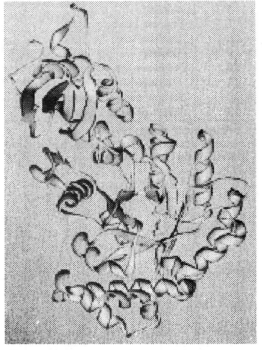

(b)

Figure 18-21. **The X-ray structure of RuBP carboxylase.**

$$3\ CO_2 + 9\ ATP + 6\ NADPH \longrightarrow GAP + 9\ ADP + 8\ P_i + 6\ NADP^+$$

Table 18-1. Standard and Physiological Free Energy Changes for the Reactions of the Calvin Cycle

Step[a]	Enzyme	$\Delta G^{\circ\prime}$ $(kJ \cdot mol^{-1})$	ΔG $(kJ \cdot mol^{-1})$
1	Phosphoribulokinase	−21.8	−15.9
2	Ribulose bisphosphate carboxylase	−35.1	−41.0
3 + 4	Phosphoglycerate kinase + glyceraldehyde-3-phosphate dehydrogenase	+18.0	−6.7
5	Triose phosphate isomerase	−7.5	−0.8
6	Aldolase	−21.8	−1.7
7	Fructose bisphosphatase	−14.2	−27.2
8	Transketolase	+6.3	−3.8
9	Aldolase	−23.4	−0.8
10	Sedoheptulose bisphosphatase	−14.2	−29.7
11	Transketolase	+0.4	−5.9
12	Phosphopentose epimerase	+0.8	−0.4
13	Ribose phosphate isomerase	+2.1	−0.4

[a]Refer to Fig. 18-20.

Source: Bassham, J.A. and Buchanan, B.B., *in* Govindjee (Ed.), *Photosynthesis,* Vol. II, p. 155, Academic Press (1982).

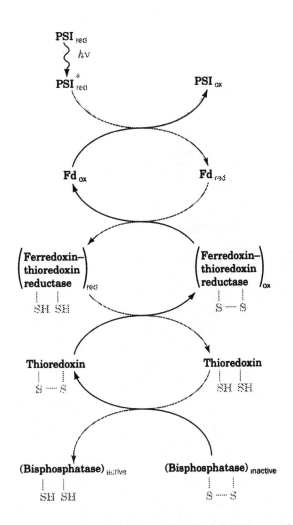

Figure 18-23. The light-activation mechanism of FBPase and SBPase.

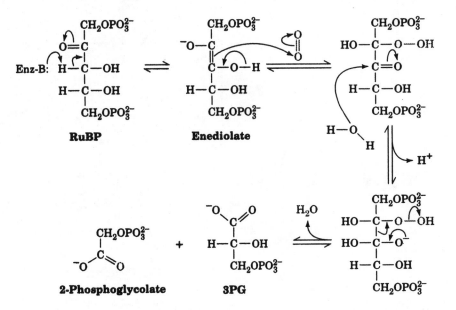

Figure 18-24. The probable mechanism of the oxygenase reaction catalyzed by RuBP carboxylase–oxygenase.

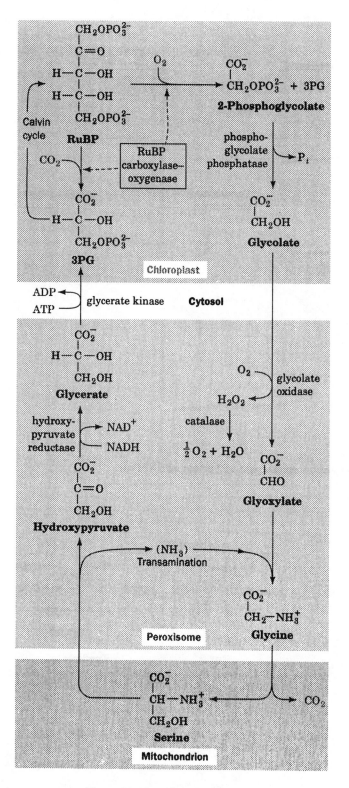

Figure 18-25. Photorespiration.

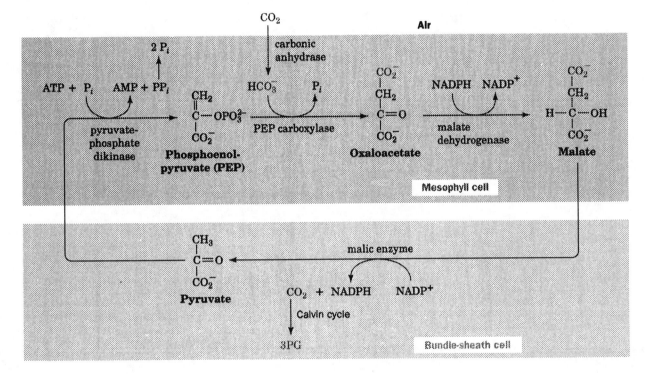

Figure 18-26. The C$_4$ pathway.

229

CHAPTER 19

LIPID METABOLISM

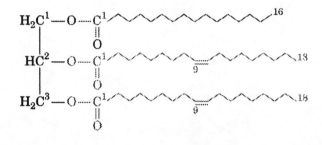

1-Palmitoyl-2,3-dioleoyl-glycerol

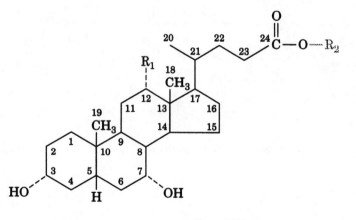

R₁ = OH | R₁ = H

	R₁ = OH	R₁ = H
R₂ = H	Cholic acid	Chenodeoxycholic acid
R₂ = NH—CH₂—COOH	Glycocholic acid	Glycochenodeoxycholic acid
R₂ = NH—CH₂—CH₂—SO₃H	Taurocholic acid	Taurochenodeoxycholic acid

Figure 19-1. Structures of the major bile acids and their glycine and taurine conjugates.

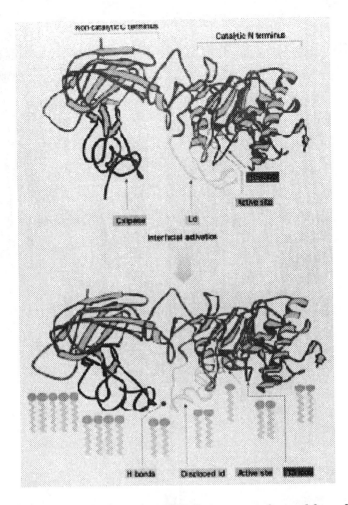

Figure 19-2. The mechanism of interfacial activation of triacylglycerol lipase.

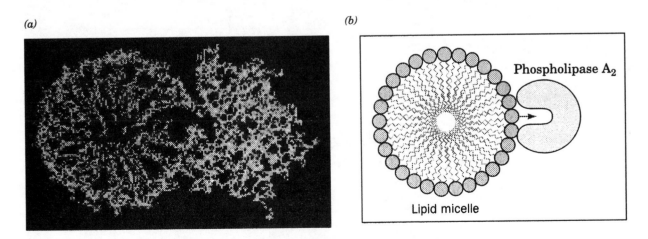

(a)

(b)

Phospholipase A₂

Lipid micelle

Figure 19-3. Substrate binding to phospholipase A₂.

231

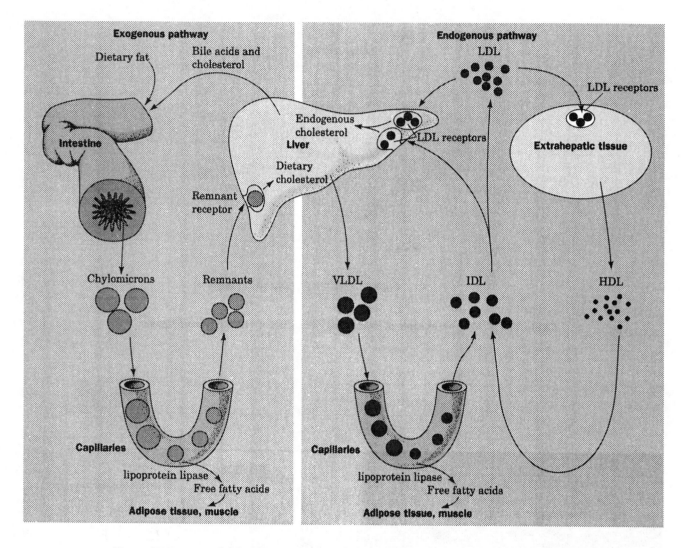

Figure 19-5. **A model for plasma triacylglycerol and cholesterol transport in humans.**

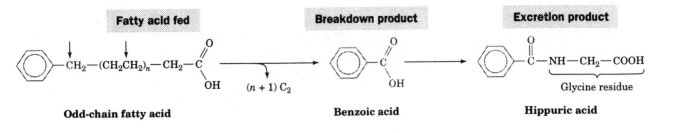

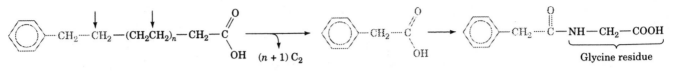

Figure 19-6. Franz Knoop's classic experiment.

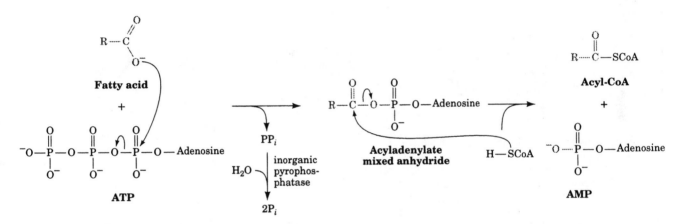

Figure 19-7. The mechanism of fatty acid activation catalyzed by acyl-CoA synthetase.

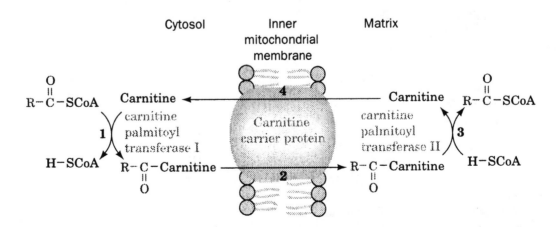

Figure 19-8. The transport of fatty acids into the mitochondrion.

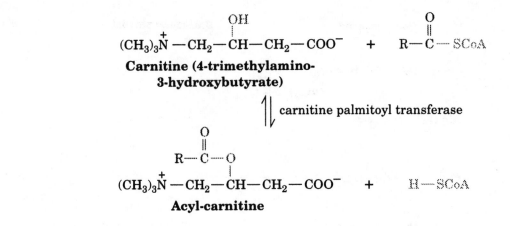

Carnitine (4-trimethylamino-3-hydroxybutyrate)

carnitine palmitoyl transferase

Acyl-carnitine

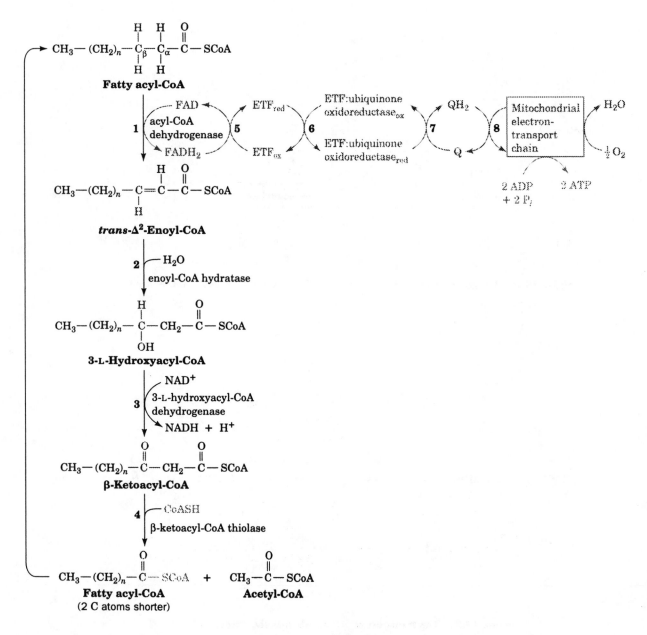

Figure 19-9. *Key to Metabolism.* **The β-oxidation pathway of fatty acyl-CoA.**

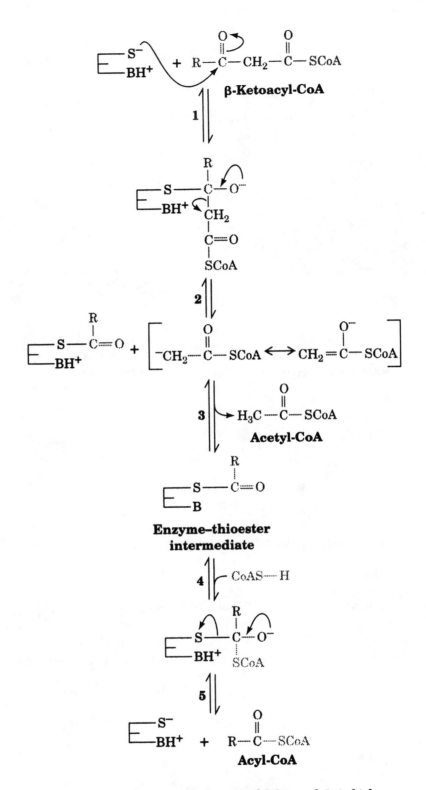

Figure 19-11. The mechanism of action of β-ketoacyl-CoA thiolase.

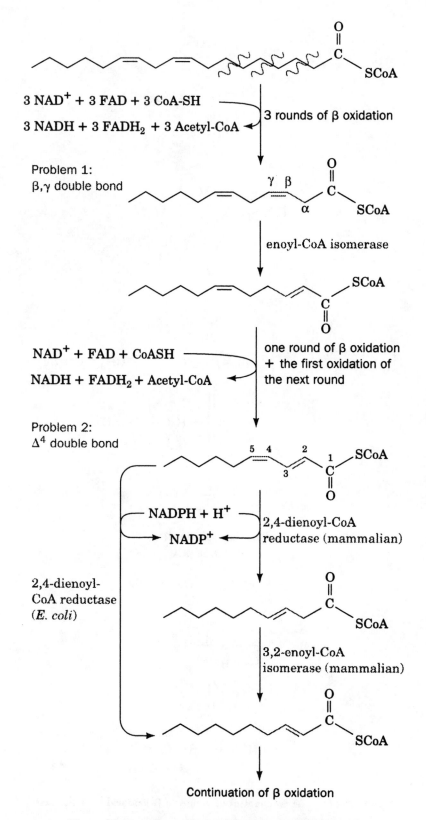

Figure 19-12. The oxidation of unsaturated fatty acids.

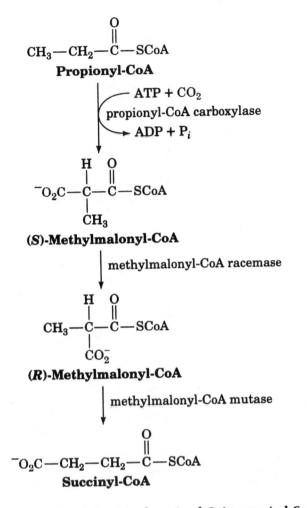

Figure 19-13. The conversion of propionyl-CoA to succinyl-CoA.

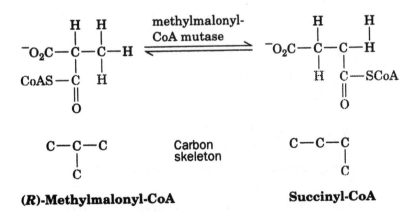

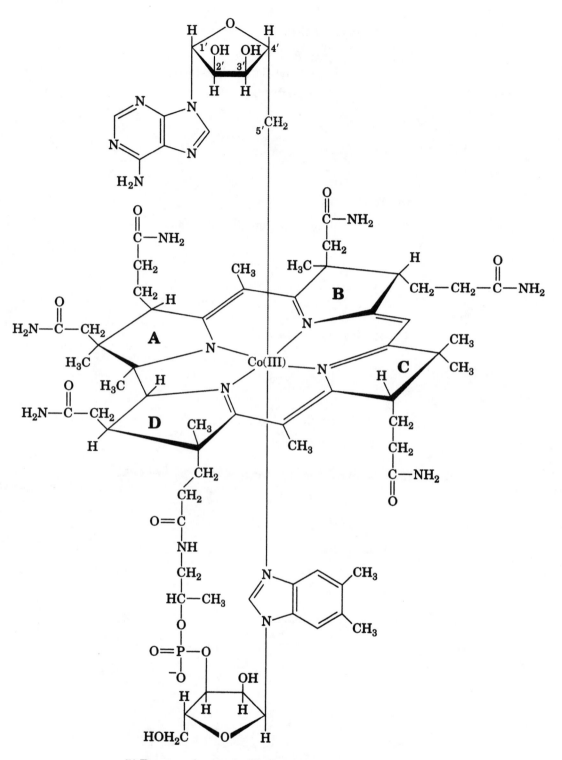

5′-Deoxyadenosylcobalamin (coenzyme B₁₂)

Figure 19-14. The structure of 5′-deoxyadenosylcobalamin.

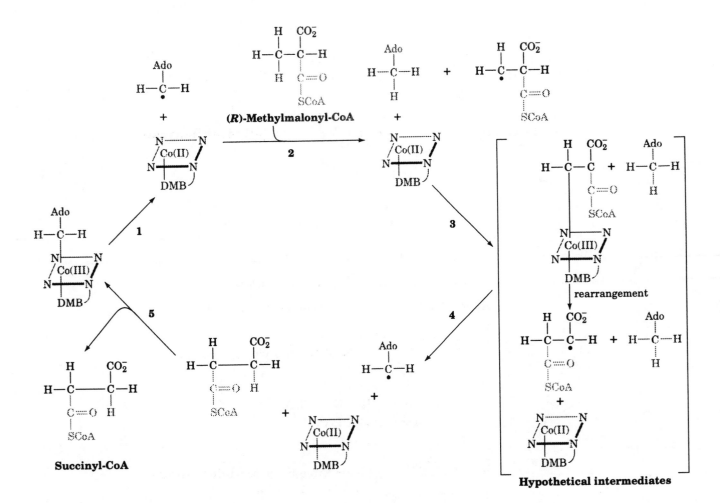

Figure 19-15. The proposed mechanism of methylmalonyl-CoA mutase.

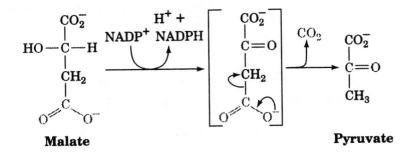

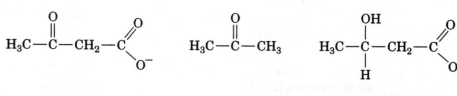

Acetoacetate   Acetone   D-β-Hydroxybutyrate

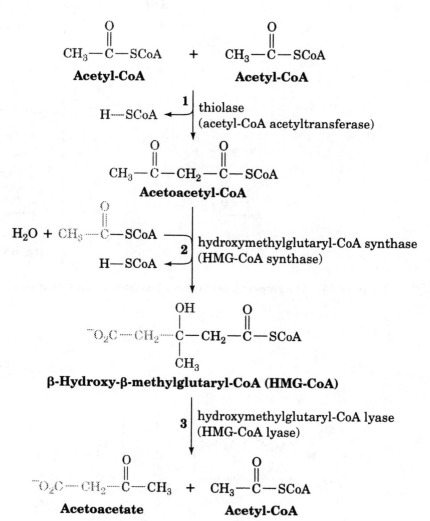

Figure 19-17.   Ketogenesis.

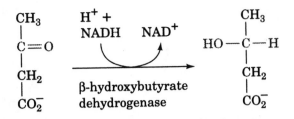

Acetoacetate   D-β-Hydroxybutyrate

Figure 19-18. **The metabolic conversion of ketone bodies to acetyl-CoA.**

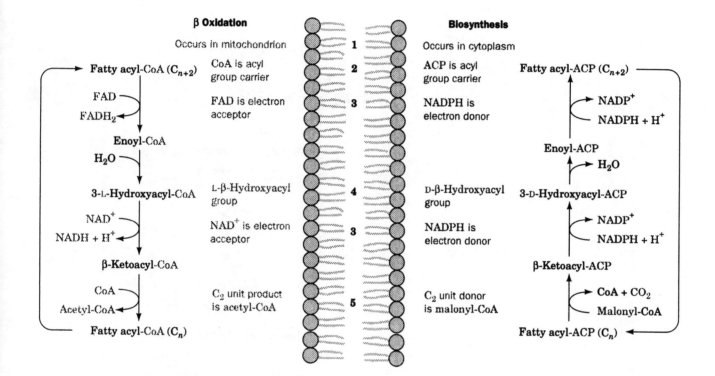

β Oxidation

Occurs in mitochondrion

Fatty acyl-CoA (C_{n+2})

FAD
FADH$_2$

Enoyl-CoA

H$_2$O

3-L-Hydroxyacyl-CoA

NAD$^+$
NADH + H$^+$

β-Ketoacyl-CoA

CoA
Acetyl-CoA

Fatty acyl-CoA (C_n)

CoA is acyl
group carrier

FAD is electron
acceptor

L-β-Hydroxyacyl
group

NAD$^+$ is electron
acceptor

C_2 unit product
is acetyl-CoA

1

2

3

4

3

5

Biosynthesis

Occurs in cytoplasm

ACP is acyl
group carrier

NADPH is
electron donor

D-β-Hydroxyacyl
group

NADPH is
electron donor

C_2 unit donor
is malonyl-CoA

Fatty acyl-ACP (C_{n+2})

NADP$^+$
NADPH + H$^+$

Enoyl-ACP

H$_2$O

3-D-Hydroxyacyl-ACP

NADP$^+$
NADPH + H$^+$

β-Ketoacyl-ACP

CoA + CO$_2$
Malonyl-CoA

Fatty acyl-ACP (C_n)

Figure 19-19. A comparison of fatty acid β oxidation and fatty acid biosynthesis.

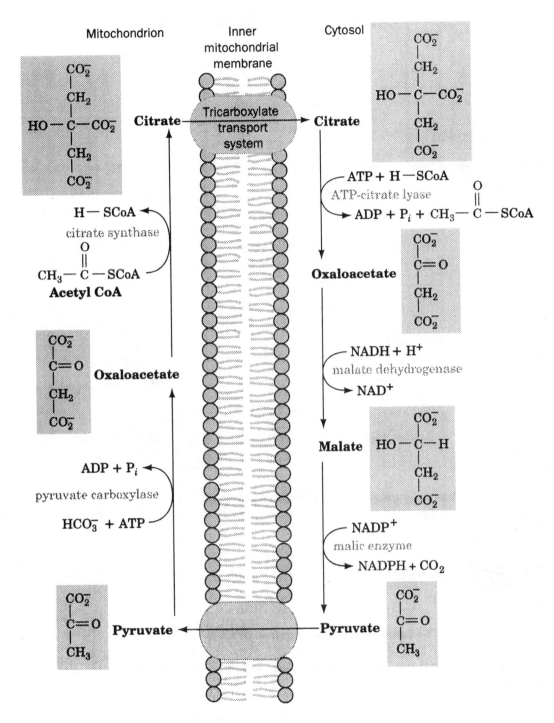

Figure 19-20. The tricarboxylate transport system.

ACETYL-CoA CARBOXYLASE

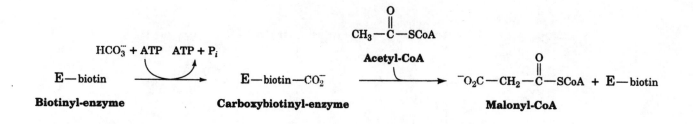

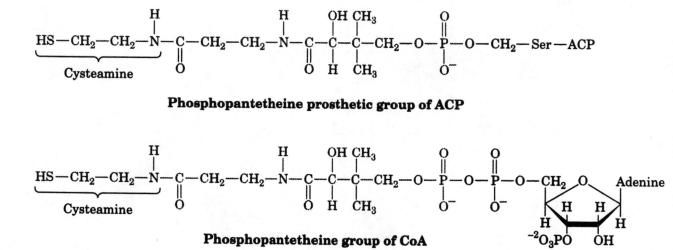

Phosphopantetheine prosthetic group of ACP

Phosphopantetheine group of CoA

Figure 19-21. **The phosphopantetheine group in acyl-carrier protein (ACP) and in CoA.**

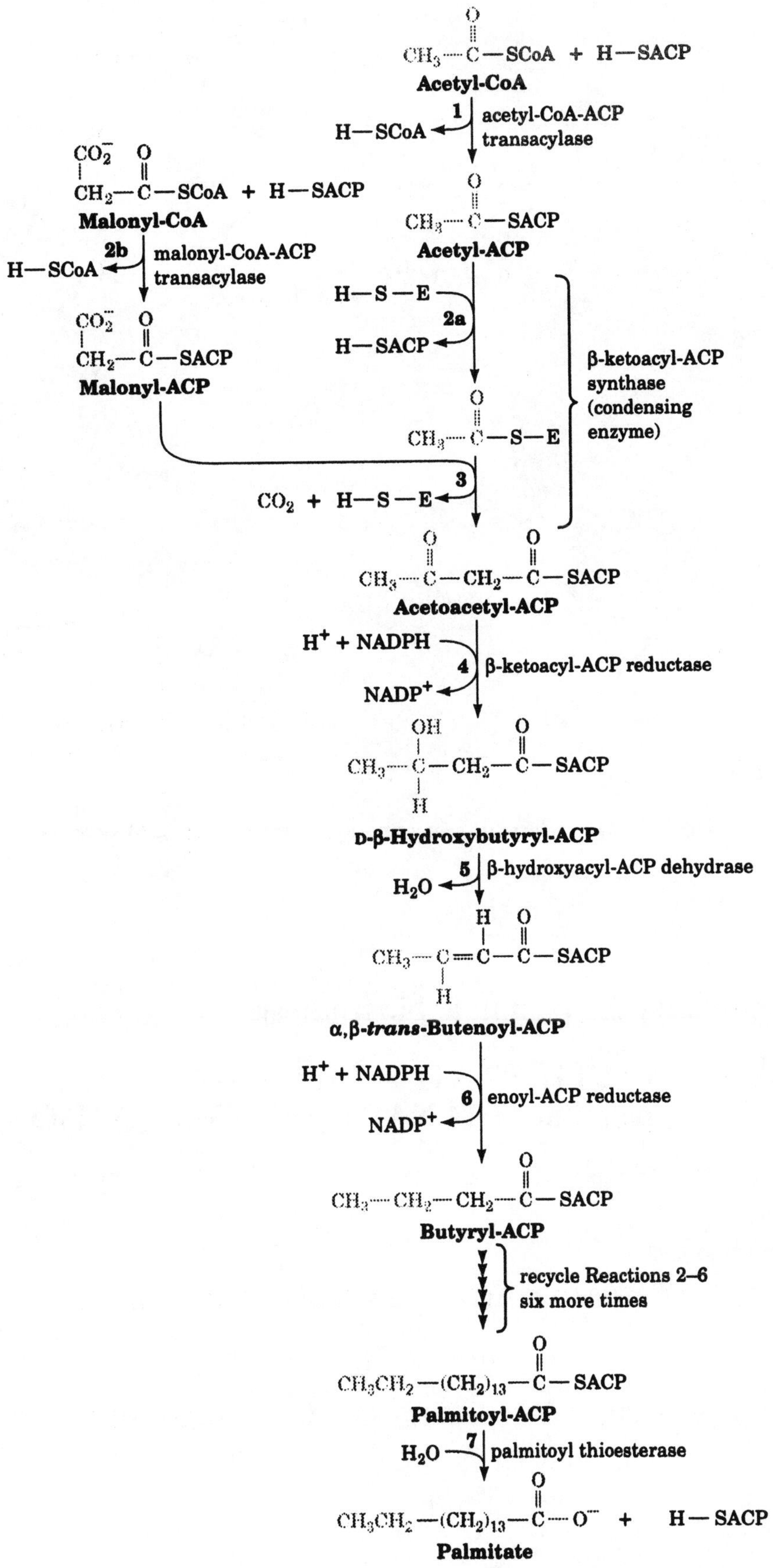

Figure 19-22. *Key to Metabolism.* The reaction sequence for the biosynthesis of fatty acids.

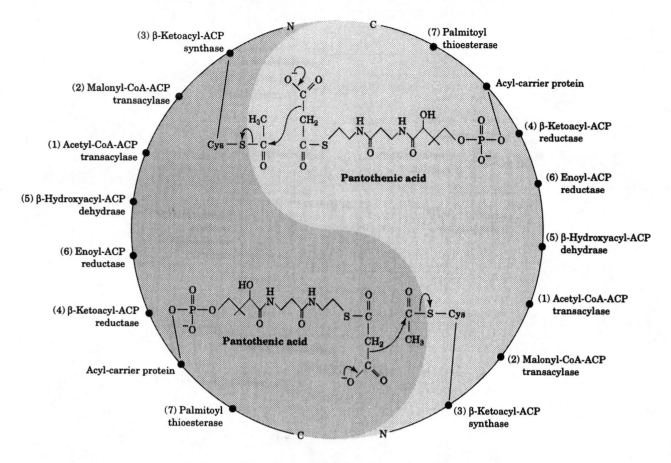

Figure 19-23. Schematic representation of animal fatty acid synthase.

Overall stoichiometry for palmitate biosynthesis:

$$8 \text{ Acetyl-CoA} + 14 \text{ NADPH} + 7 \text{ ATP} + 7 \text{ H}^+ \longrightarrow$$
$$\text{palmitate} + 14 \text{ NADP}^+ + 8 \text{ CoA} + 6 \text{ H}_2\text{O} + 7 \text{ ADP} + 7 \text{ P}_i$$

FATTY ACYL-CoA DESATURASES

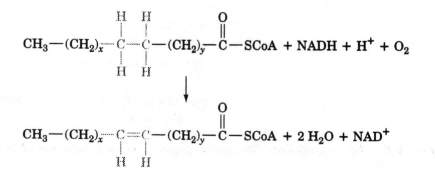

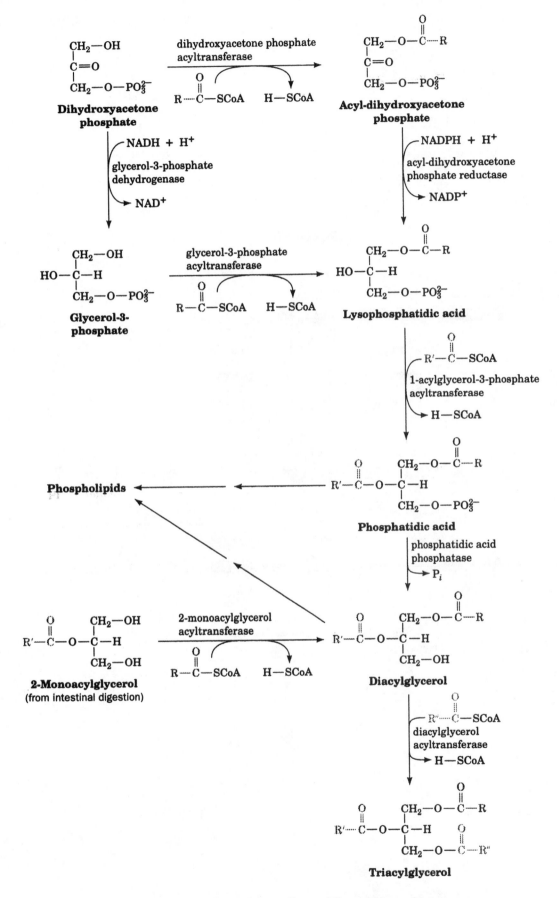

Figure 19-25. The reactions of triacylglycerol biosynthesis.

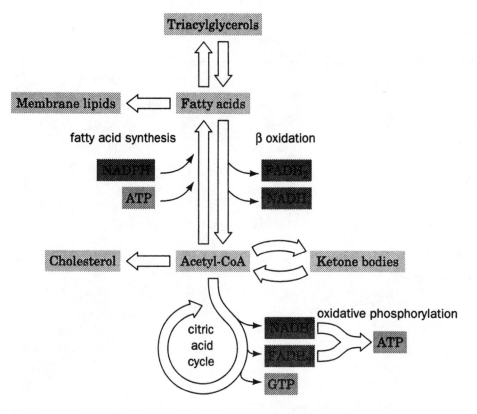

Figure 19-26. **A summary of lipid metabolism.**

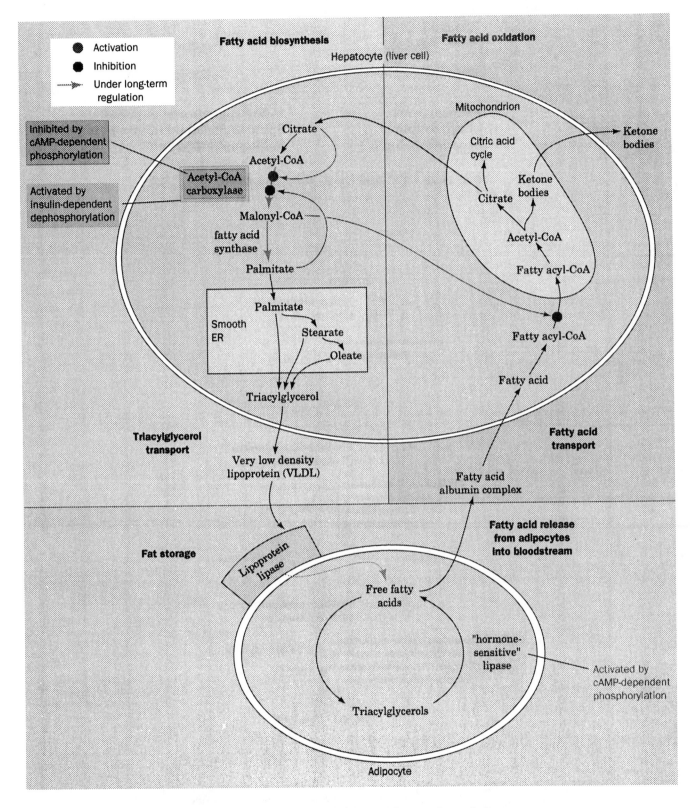

Figure 19-27. Sites of regulation of fatty acid metabolism.

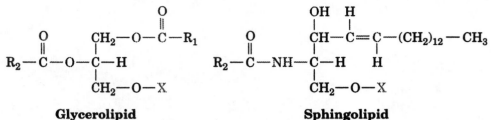

X = H	**1,2-Diacylglycerol**	***N*-Acylsphingosine (ceramide)**
X = Carbohydrate	**Glyceroglycolipid**	**Sphingoglycolipid (glycosphingolipid)**
X = Phosphate ester	**Glycerophospholipid**	**Sphingophospholipid**

Figure 19-28. The glycerolipids and sphingolipids.

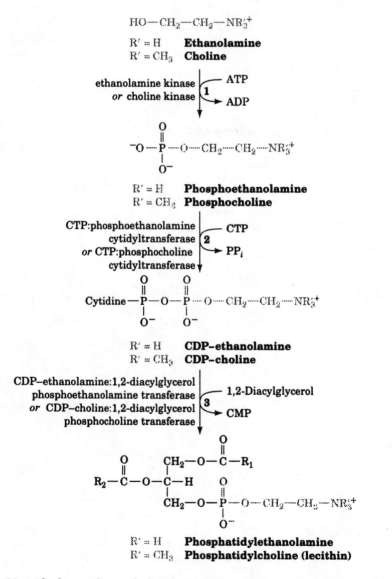

Figure 19-29. The biosynthesis of phosphatidylethanolamine and phosphatidylcholine.

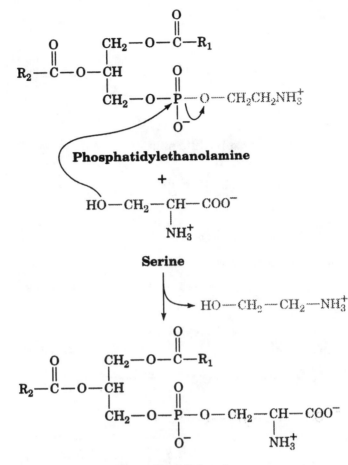

Phosphatidylethanolamine

+

Serine

Phosphatidylserine

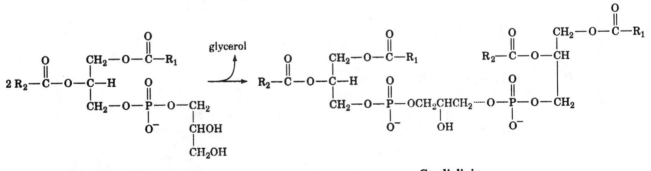

Phosphatidylglycerol   **Cardiolipin**

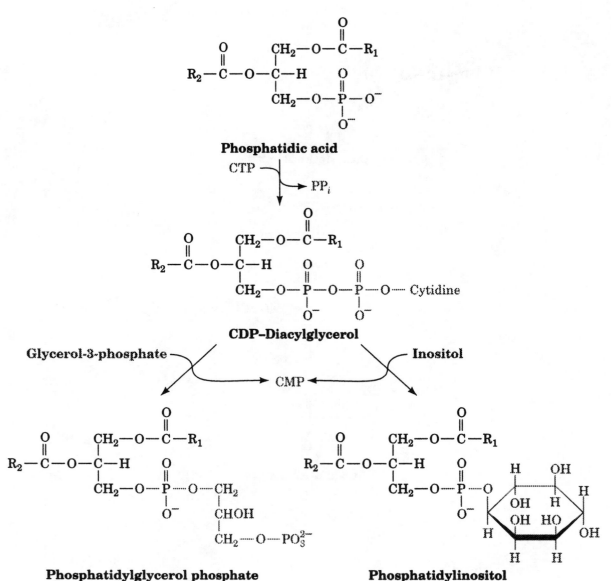

Phosphatidic acid

CDP–Diacylglycerol

Phosphatidylglycerol phosphate

Phosphatidylinositol

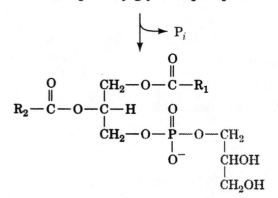

Phosphatidylglycerol

Figure 19-30. **The biosynthesis of phosphatidylinositol and phosphatidylglycerol.**

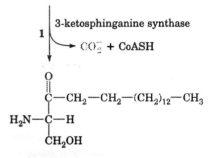

Palmitoyl-CoA   **Serine**

1   3-ketosphinganine synthase

CO_2 + CoASH

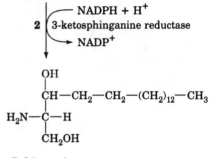

3-Ketosphinganine
(3-ketodihydrosphingosine)

2   NADPH + H^+
3-ketosphinganine reductase
$NADP^+$

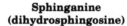

Sphinganine
(dihydrosphingosine)

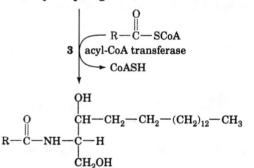

3    $R-\overset{O}{\overset{\|}{C}}-SCoA$
acyl-CoA transferase
CoASH

Dihydroceramide
(*N*-acylsphinganine)

4   FAD
dihydroceramide reductase
$FADH_2$

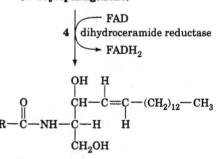

Ceramide
(*N*-acylsphingosine)

Figure 19-31. The biosynthesis of ceramide (*N*-acylsphingosine).

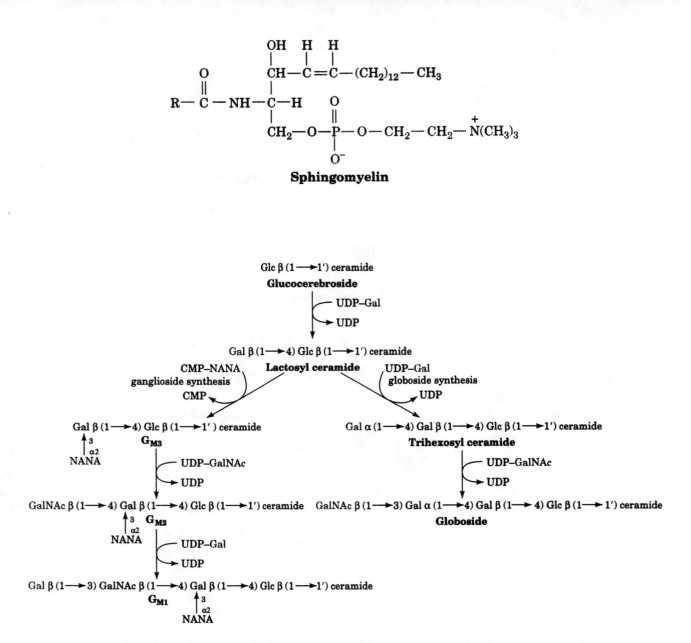

Sphingomyelin

Glc β (1 ⟶ 1′) ceramide
Glucocerebroside

UDP–Gal
UDP

Gal β (1 ⟶ 4) Glc β (1 ⟶ 1′) ceramide
Lactosyl ceramide

CMP–NANA
ganglioside synthesis
CMP

UDP–Gal
globoside synthesis
UDP

Gal β (1 ⟶ 4) Glc β (1 ⟶ 1′) ceramide
G_M3

3
α2
NANA

UDP–GalNAc
UDP

Gal α (1 ⟶ 4) Gal β (1 ⟶ 4) Glc β (1 ⟶ 1′) ceramide
Trihexosyl ceramide

UDP–GalNAc
UDP

GalNAc β (1 ⟶ 4) Gal β (1 ⟶ 4) Glc β (1 ⟶ 1′) ceramide
G_M2

3
α2
NANA

UDP–Gal
UDP

GalNAc β (1 ⟶ 3) Gal α (1 ⟶ 4) Gal β (1 ⟶ 4) Glc β (1 ⟶ 1′) ceramide
Globoside

Gal β (1 ⟶ 3) GalNAc β (1 ⟶ 4) Gal β (1 ⟶ 4) Glc β (1 ⟶ 1′) ceramide
G_M1

3
α2
NANA

Figure 19-32. **The biosynthesis of globosides and G_M gangliosides.**

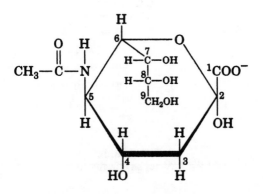

***N*-Acetylneuraminic acid**
(NANA, sialic acid)

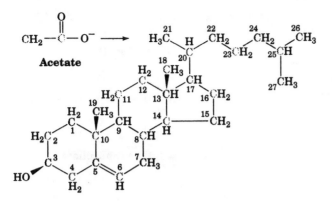

Acetate

CHOLESTEROL

Isoprene
(2-methyl-1,3-butadiene)

An isoprene unit

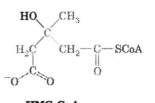

HMG-CoA

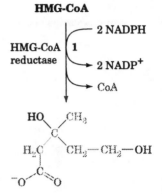

HMG-CoA
reductase **1**

2 NADPH

2 NADP$^+$

CoA

Mevalonate

mevalonate-5-
phosphotransferase **2**

ATP

ADP

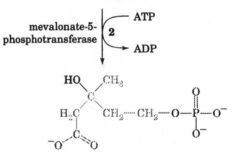

Phosphomevalonate

phosphomevalonate
kinase **3**

ATP

ADP

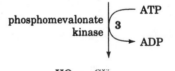

5-Pyrophosphomevalonate

pyrophospho-
mevalonate
decarboxylase **4**

ATP

ADP + P$_i$ + CO$_2$

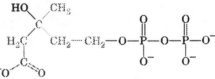

Isopentenyl pyrophosphate

Figure 19-33. **The formation of isopentenyl**
pyrophosphate from HMG-CoA.

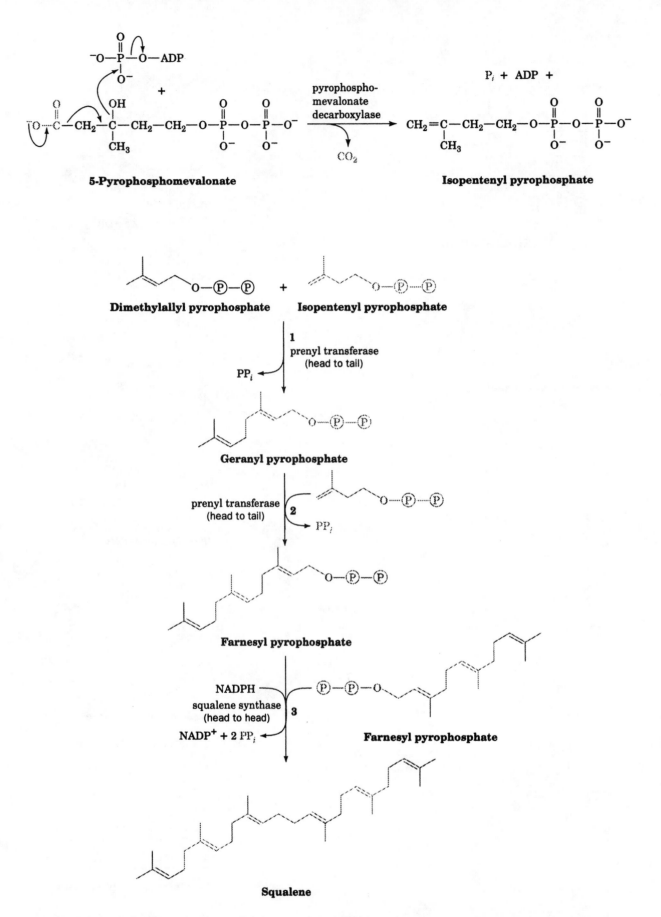

Figure 19-34. The formation of squalene from isopentenyl pyrophosphate and dimethylallyl pyrophosphate.

PRENYL TRANSFERASE REACTION MECHANISM

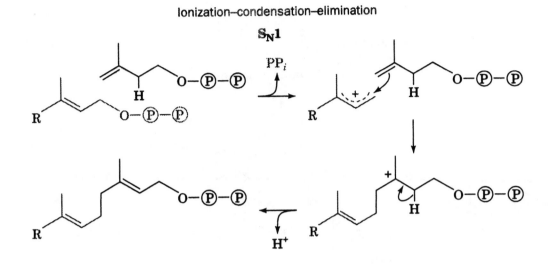

Ionization–condensation–elimination

S_N1

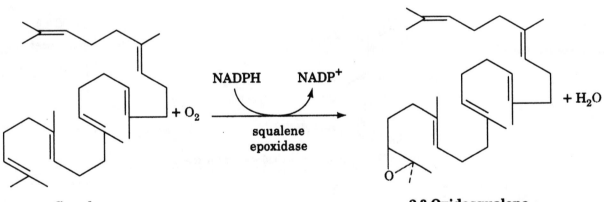

Squalene

2,3-Oxidosqualene

Figure 19-35. The squalene epoxidase reaction.

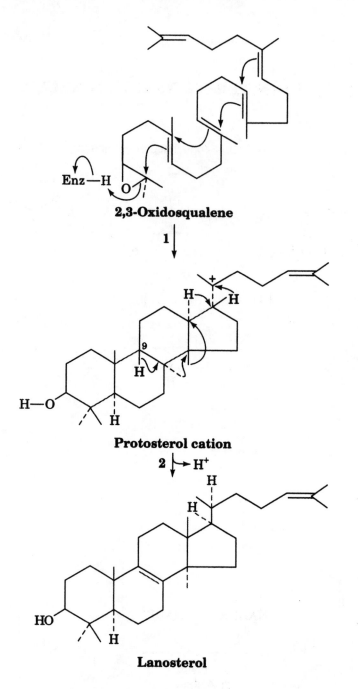

2,3-Oxidosqualene

Protosterol cation

Lanosterol

Figure 19-36. **The squalene oxidocyclase reaction.**

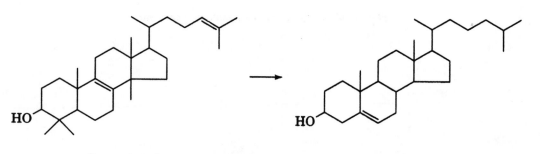

Lanosterol

Cholesterol

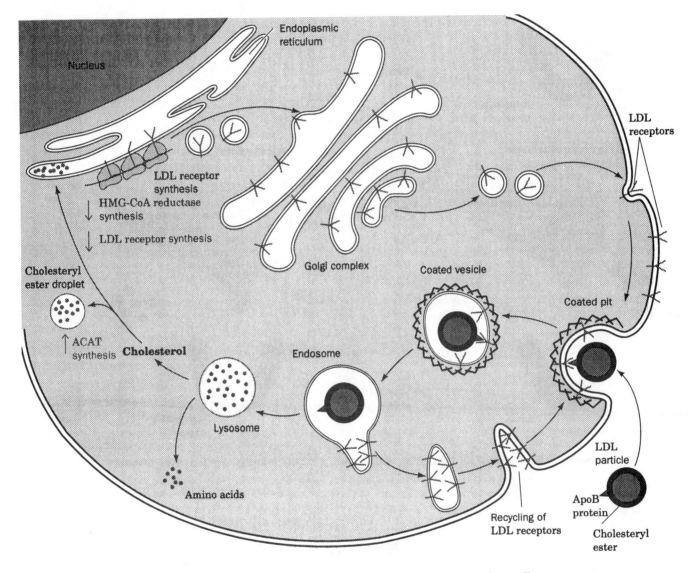

Figure 19-37. LDL receptor-mediated endocytosis in mammalian cells.

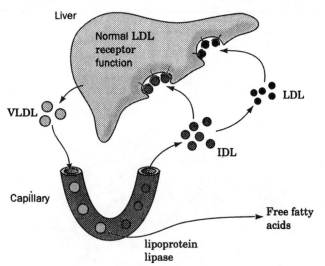

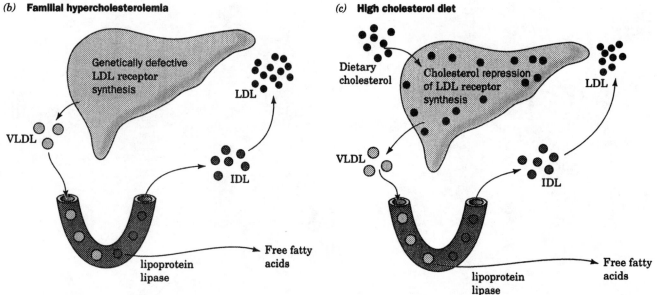

Figure 19-38. Uptake of plasma LDL and control of LDL production by liver LDL receptors.

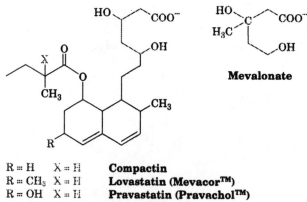

R = H   X = H   **Compactin**
R = CH₃   X = H   **Lovastatin (Mevacor™)**
R = OH   X = H   **Pravastatin (Pravachol™)**
R = CH₃   X = CH₃   **Simvastatin (Zocor™)**

AMINO ACID METABOLISM

Table 20-1. **Half-Lives of Some Rat Liver Enzymes**

Enzyme	Half-Life (h)
Short-Lived Enzymes	
Ornithine decarboxylase	0.2
RNA polymerase I	1.3
Tyrosine aminotransferase	2.0
Serine dehydratase	4.0
PEP carboxylase	5.0
Long-Lived Enzymes	
Aldolase	118
GAPDH	130
Cytochrome *b*	130
LDH	130
Cytochrome *c*	150

Source: Dice, J.F. and Goldberg, A.L., *Arch. Biochem. Biophys.* 170, 214 (1975).

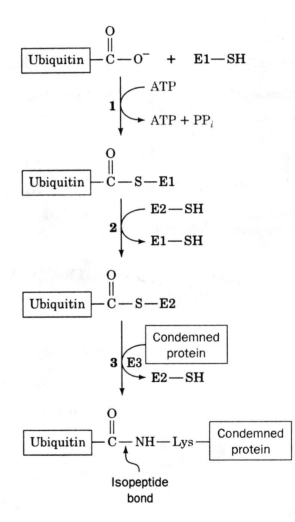

Figure 20-2.   The reactions involved in protein ubiquitination.

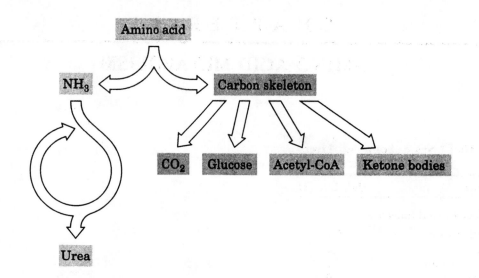

Figure 20-5. Overview of amino acid catabolism.

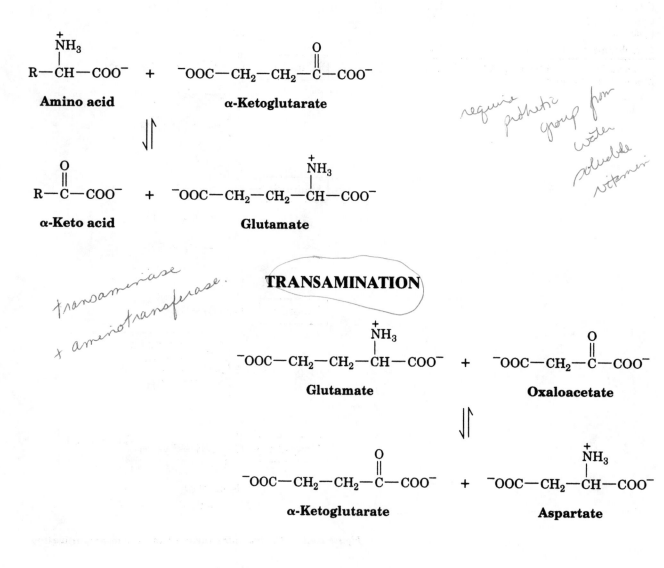

require prosthetic group from water soluble vitamin

transaminase + aminotransferase.

TRANSAMINATION

262

(a)

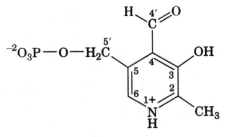

**Pyridoxal-5′-
phosphate (PLP)**

(b)

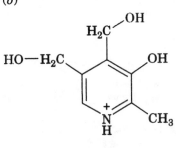

**Pyridoxine
(vitamin B₆)**

(c)

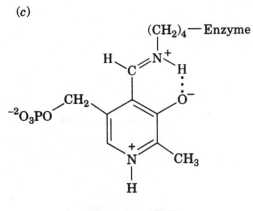

**Enzyme–PLP
Schiff base**

(d)

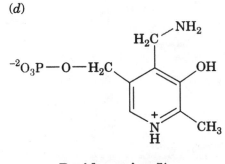

**Pyridoxamine-5′-
phosphate (PMP)**

Figure 20-6. Forms of pyridoxal-5′-phosphate.

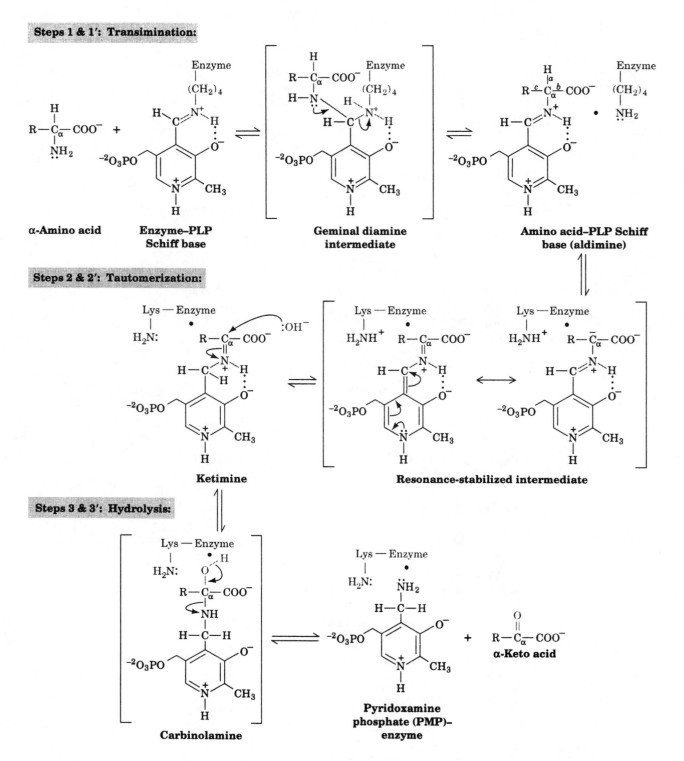

Steps 1 & 1': Transimination:

α-Amino acid   Enzyme–PLP Schiff base   Geminal diamine intermediate   Amino acid–PLP Schiff base (aldimine)

Steps 2 & 2': Tautomerization:

Ketimine   Resonance-stabilized intermediate

Steps 3 & 3': Hydrolysis:

Carbinolamine   Pyridoxamine phosphate (PMP)–enzyme   α-Keto acid

Figure 20-7. *Key to Function.* **The mechanism of PLP-dependent enzyme-catalyzed transamination.**

264

GLUTAMATE DEHYDROGENASE

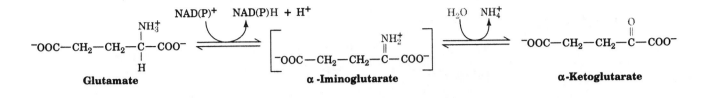

Glutamate   α-Iminoglutarate   α-Ketoglutarate

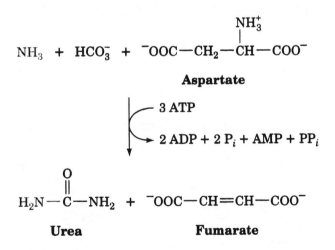

NH_3 $H_2N-\overset{\displaystyle O}{\overset{\|}{C}}-NH_2$

Ammonia   **Urea**   **Uric acid**

THE UREA CYCLE

$$NH_3 + HCO_3^- + {}^-OOC-CH_2-\overset{\displaystyle NH_3^+}{\overset{|}{CH}}-COO^-$$

Aspartate

3 ATP

2 ADP + 2 P$_i$ + AMP + PP$_i$

$$H_2N-\overset{\displaystyle O}{\overset{\|}{C}}-NH_2 + {}^-OOC-CH=CH-COO^-$$

Urea   **Fumarate**

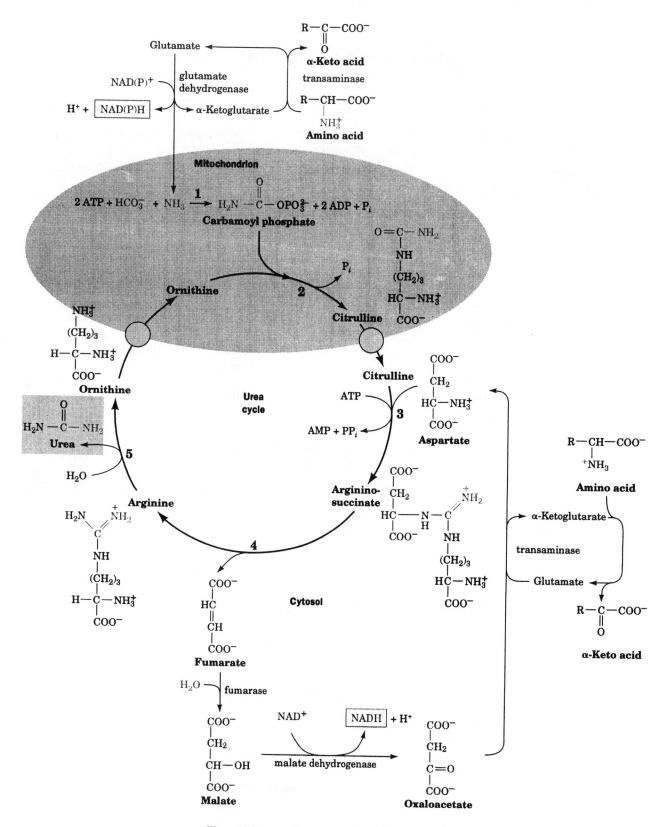

Figure 20-8. **Key to Metabolism. The urea cycle.**

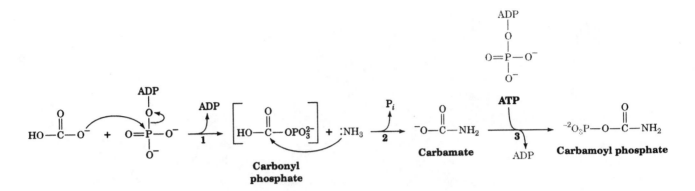

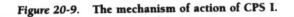

Figure 20-9. The mechanism of action of CPS I.

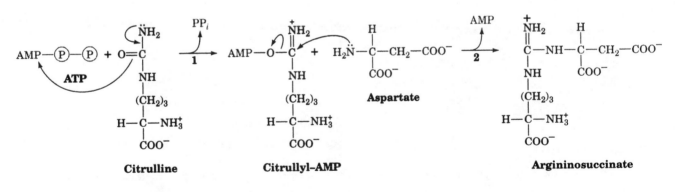

Figure 20-10. The mechanism of action of argininosuccinate synthetase.

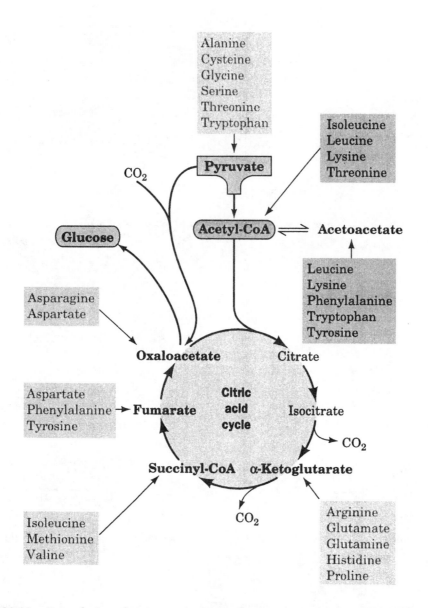

Figure 20-11. Degradation of amino acids to one of seven common metabolic intermediates.

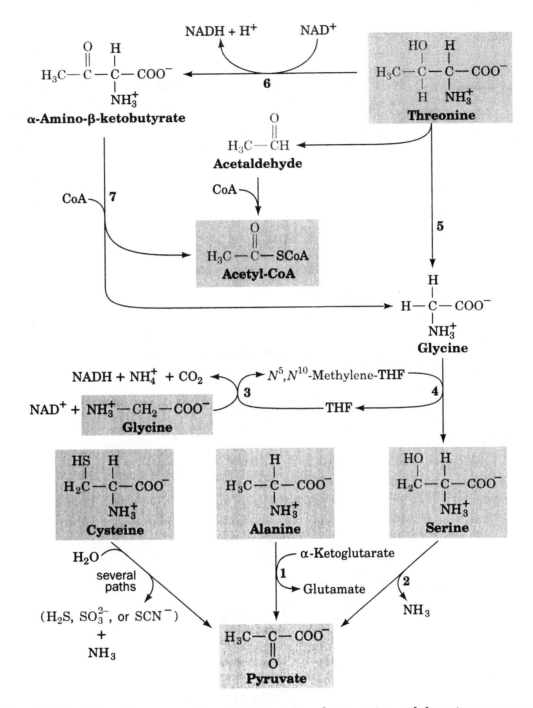

Figure 20-12. The pathways converting alanine, cysteine, glycine, serine, and threonine to pyruvate.

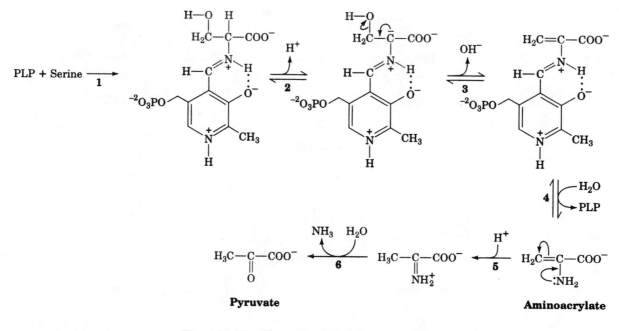

PLP + Serine $\xrightarrow{\text{1}}$

$\xrightarrow{\text{2}}$ H$^+$

$\xrightarrow{\text{3}}$ OH$^-$

$\xleftarrow{\text{4}}$ H$_2$O, PLP

Pyruvate $\xleftarrow{\text{6}}$ NH$_3$, H$_2$O $\quad$ H$_3$C—C—COO$^-$ $\xleftarrow{\text{5}}$ H$^+$ $\quad$ **Aminoacrylate**

Figure 20-13. **The serine dehydratase reaction.**

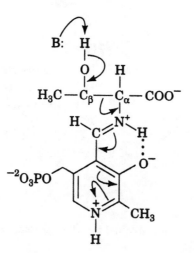

Serine Hydroxymethyltransferase Cleavage of threonine

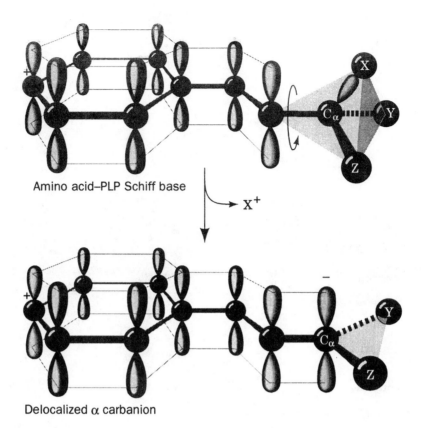

Amino acid–PLP Schiff base

$\longrightarrow X^+$

Delocalized α carbanion

Figure 20-14. The π-orbital framework of a PLP–amino acid Schiff base.

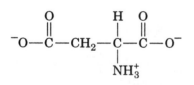

Aspartate

α-Ketoglutarate ⟶

Glutamate ⟵

aminotransferase

Oxaloacetate

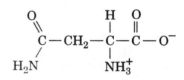

Asparagine

H_2O

NH_4^+

L-asparaginase

Aspartate

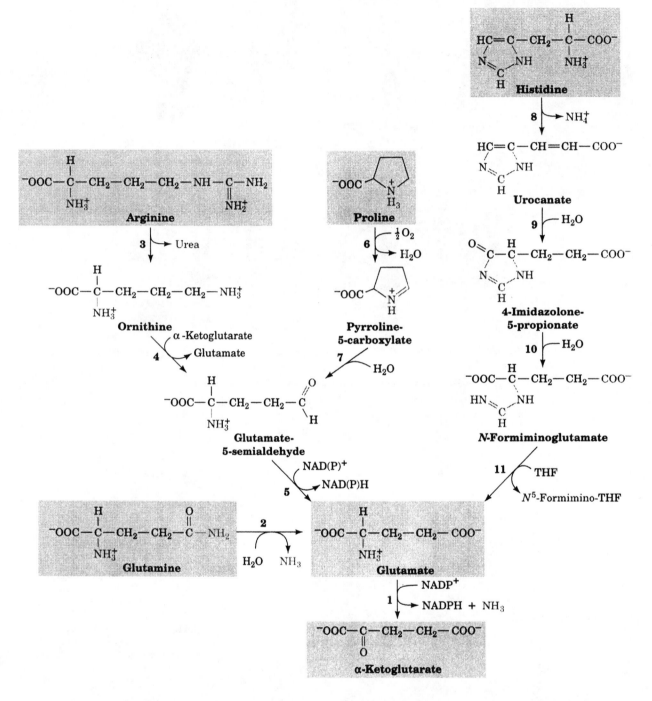

Figure 20-15. The degradation of arginine, glutamate, glutamine, histidine, and proline to α-ketoglutarate.

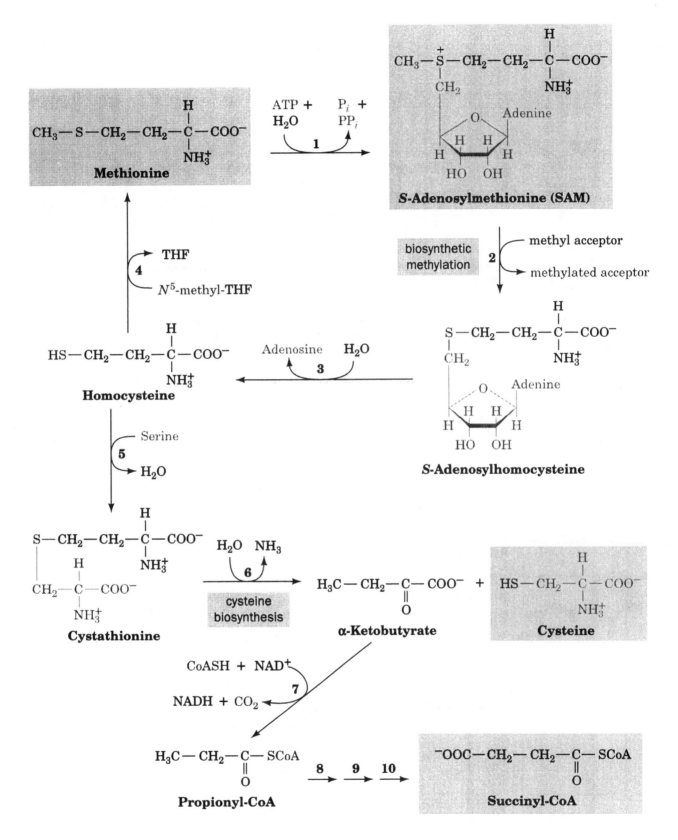

Figure 20-16. Methionine degradation.

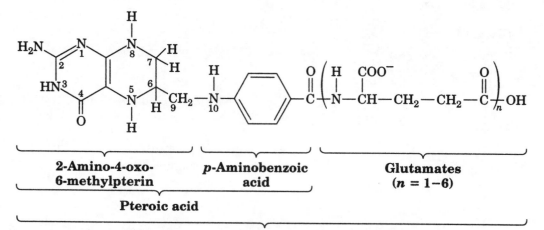

2-Amino-4-oxo-6-methylpterin

p-Aminobenzoic acid

Glutamates (n = 1–6)

Pteroic acid

Pteroylglutamic acid (tetrahydrofolate; THF)

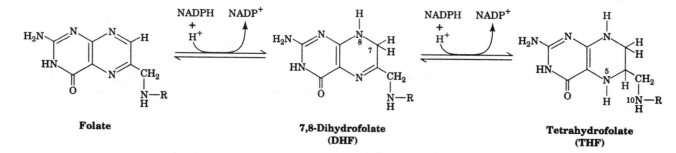

Folate

7,8-Dihydrofolate (DHF)

Tetrahydrofolate (THF)

Figure 20-17. The two-stage reduction of folate to THF.

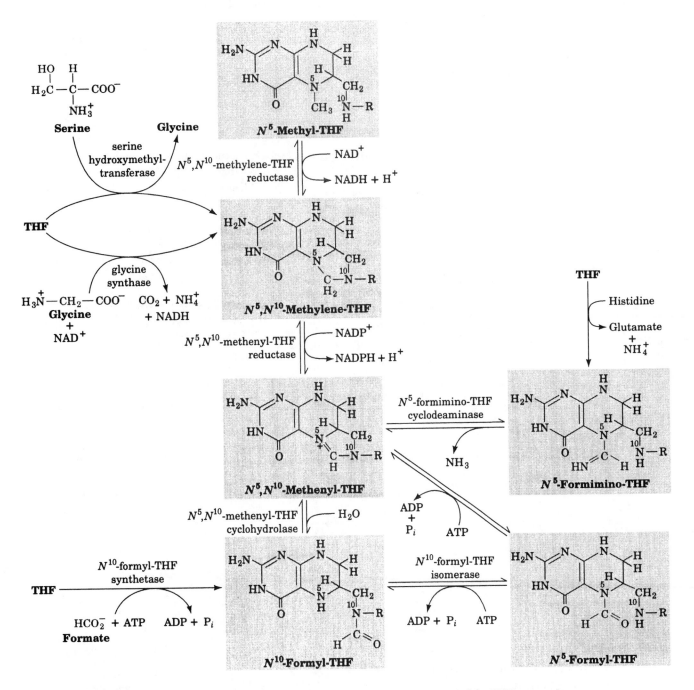

Figure 20-18. Interconversion of the C_1 units carried by THF.

Table 20-2. Oxidation Levels of C_1 Groups Carried by THF

Oxidation Level	Group Carried	THF Derivative(s)
Methanol	Methyl ($-CH_3$)	N^5-Methyl-THF
Formaldehyde	Methylene ($-CH_2-$)	N^5,N^{10}-Methylene-THF
Formate	Formyl ($-CH=O$)	N^5-Formyl-THF, N^{10}-formyl-
THF		
	Formimino ($-CH=NH$)	N^5-Formimino-THF
	Methenyl ($-CH=$)	N^5,N^{10}-Methenyl-THF

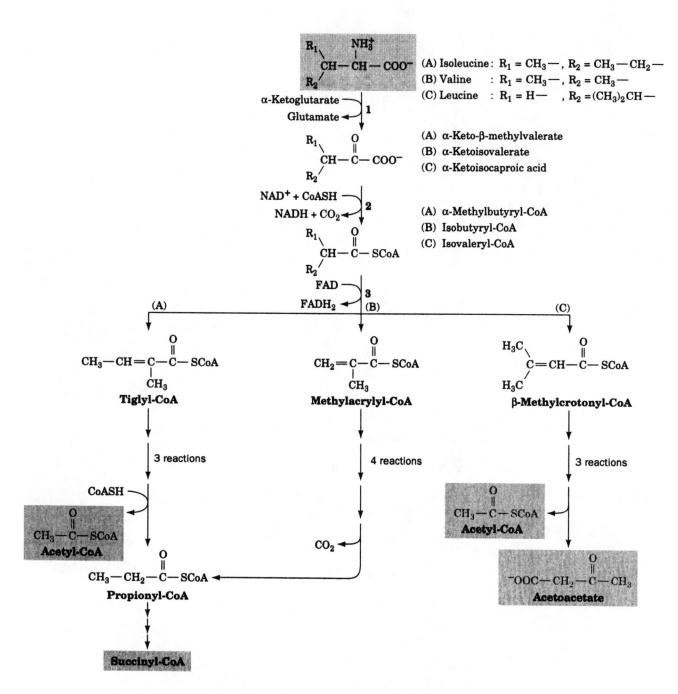

Figure 20-19. The degradation of the branched-chain amino acids.

276

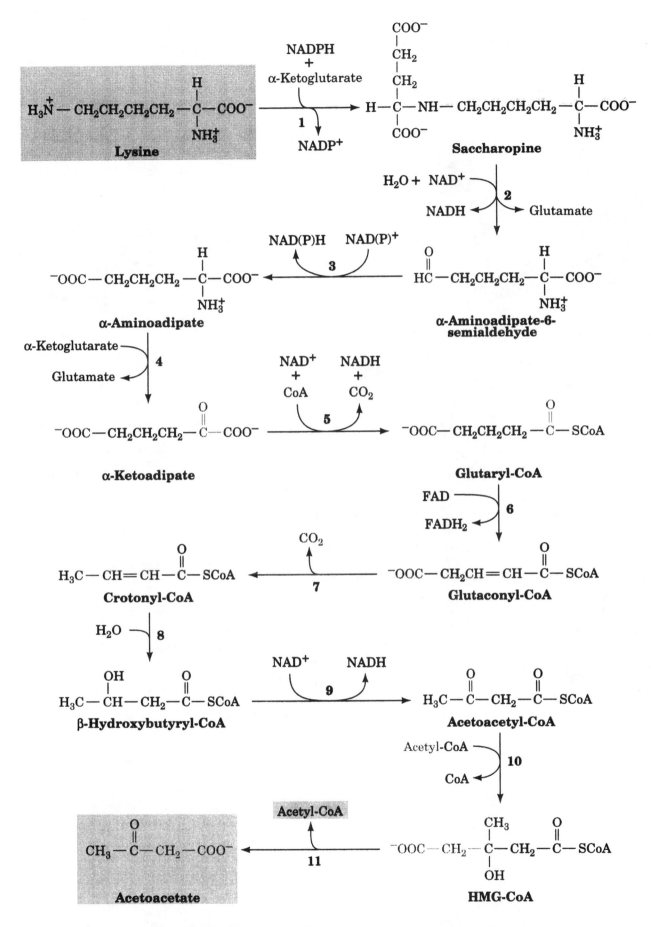

Figure 20-20. The pathway of lysine degradation in mammalian liver.

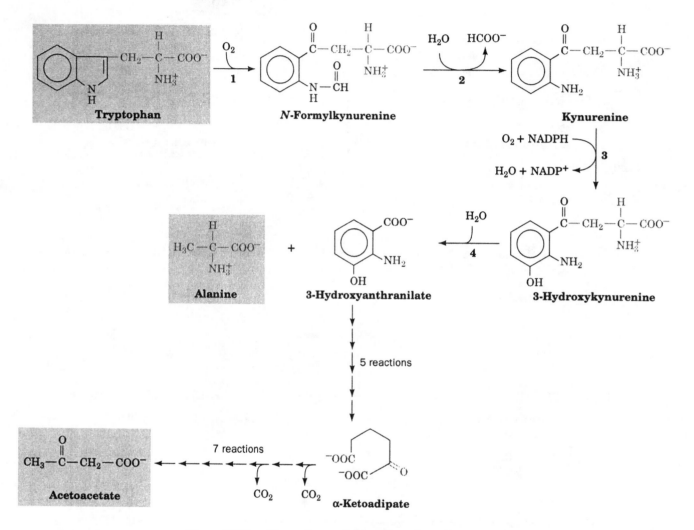

Figure 20-21. The pathway of tryptophan degradation.

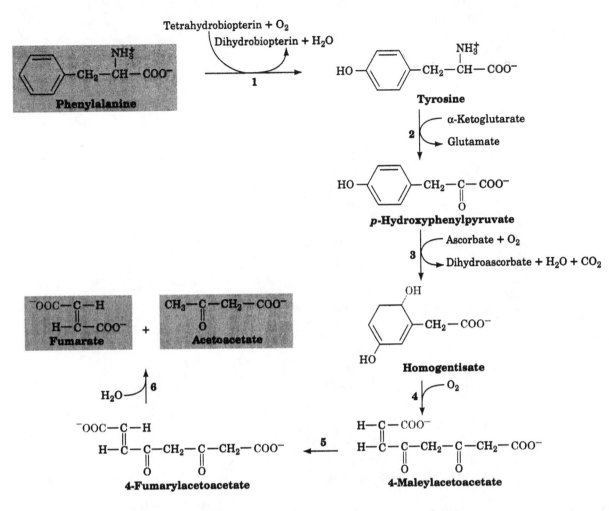

Figure 20-22. The pathway of phenylalanine degradation.

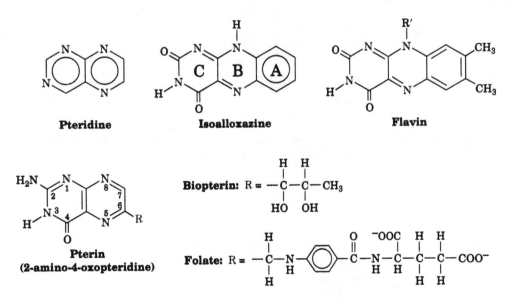

Figure 20-23. The pteridine ring nucleus of biopterin and folate.

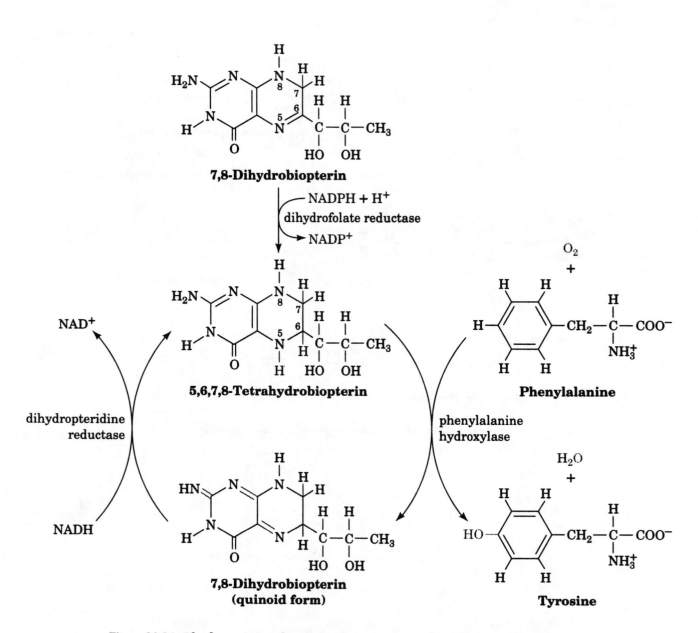

Figure 20-24. The formation, utilization, and regeneration of 5,6,7,8-tetra-hydrobiopterin in the phenylalanine hydroxylase reaction.

Table 20-3. Essential and Nonessential Amino Acids in Humans

Essential	Nonessential
Arginine[a]	Alanine
Histidine	Asparagine
Isoleucine	Aspartate
Leucine	Cysteine
Lysine	Glutamate
Methionine	Glutamine
Phenylalanine	Glycine
Threonine	Proline
Tryptophan	Serine
Valine	Tyrosine

[a]Although mammals synthesize arginine, they cleave most of it to form urea (Section 20-3A).

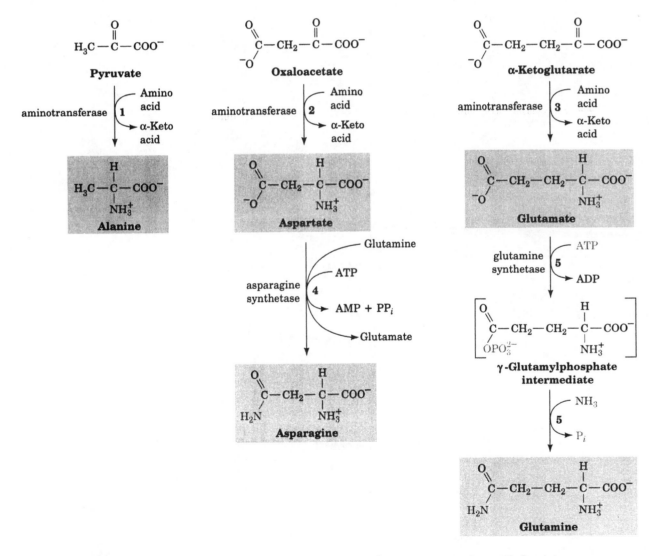

Figure 20-25. The syntheses of alanine, aspartate, glutamate, asparagine, and glutamine.

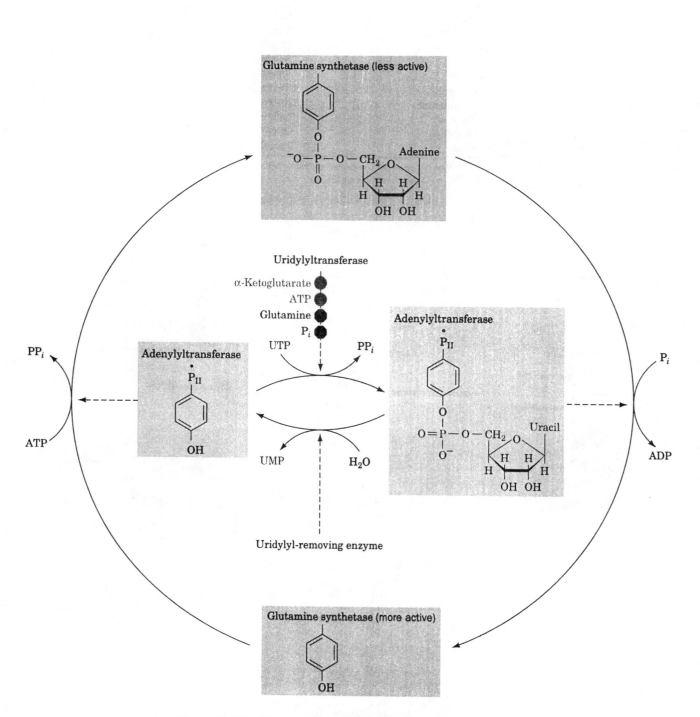

Figure 20-27. The regulation of bacterial glutamine synthetase.

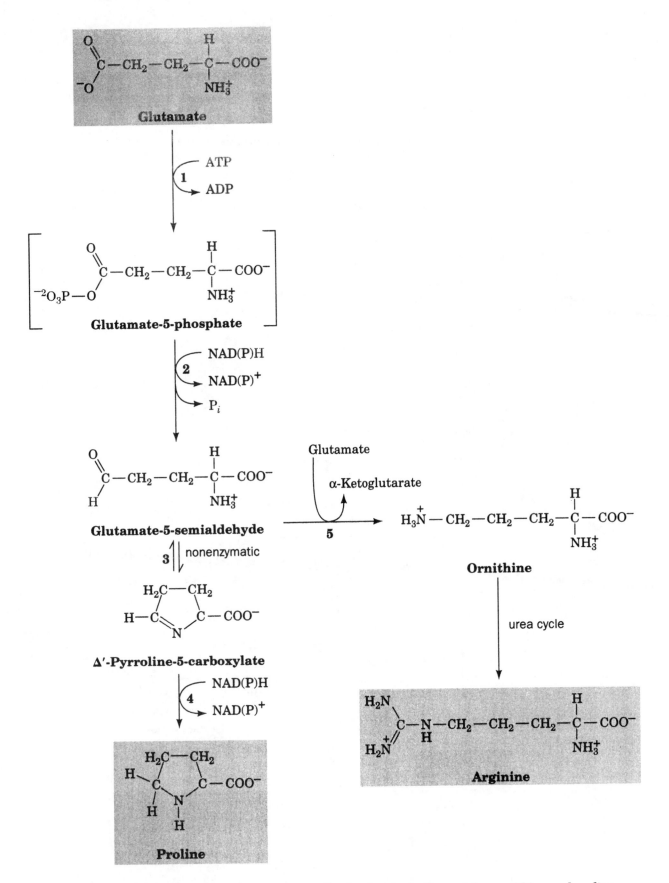

Figure 20-28. The biosynthesis of the glutamate family of amino acids: arginine, ornithine, and proline.

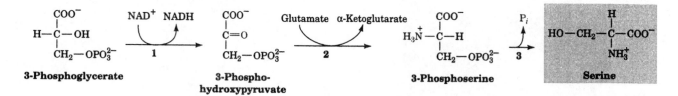

Figure 20-29. The conversion of 3-phosphoglycerate to serine.

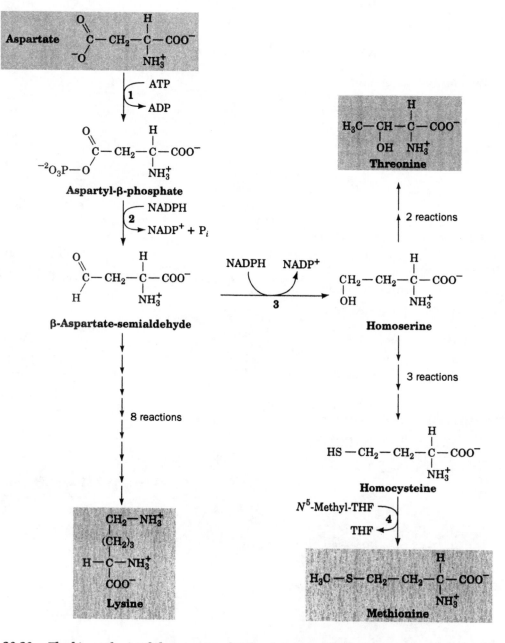

Figure 20-30. The biosynthesis of the aspartate family of amino acids: lysine, methionine, and threonine.

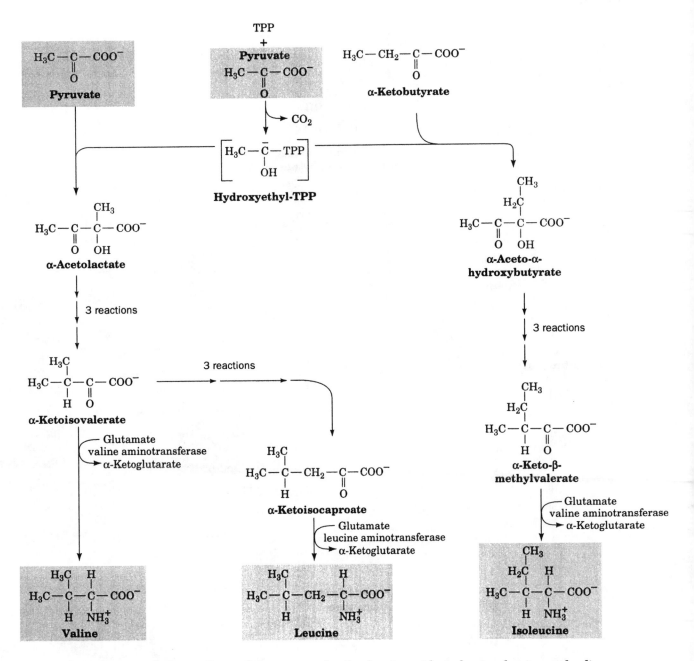

Figure 20-31. The biosynthesis of the pyruvate family of amino acids: isoleucine, leucine, and valine.

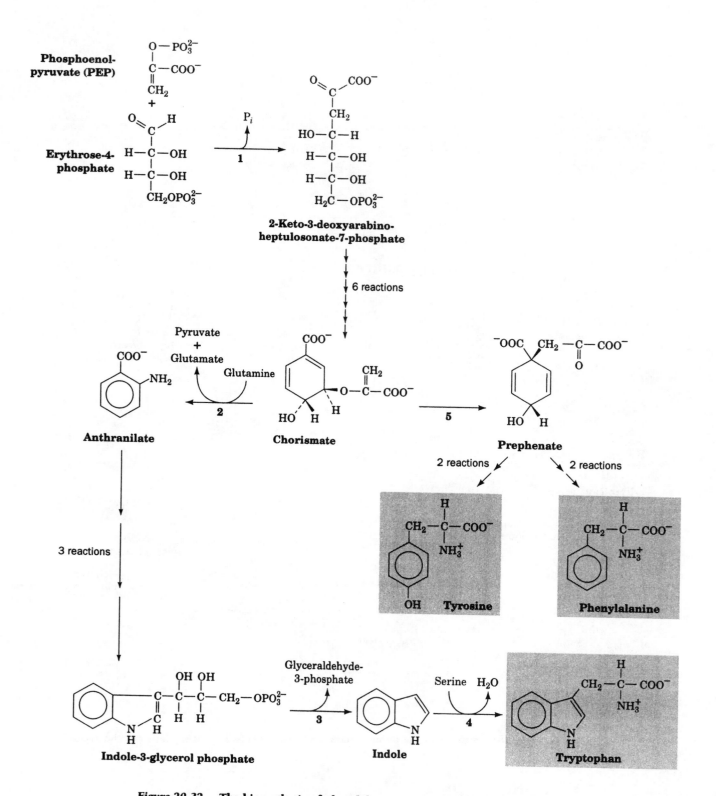

Figure 20-32. The biosynthesis of phenylalanine, tryptophan, and tyrosine.

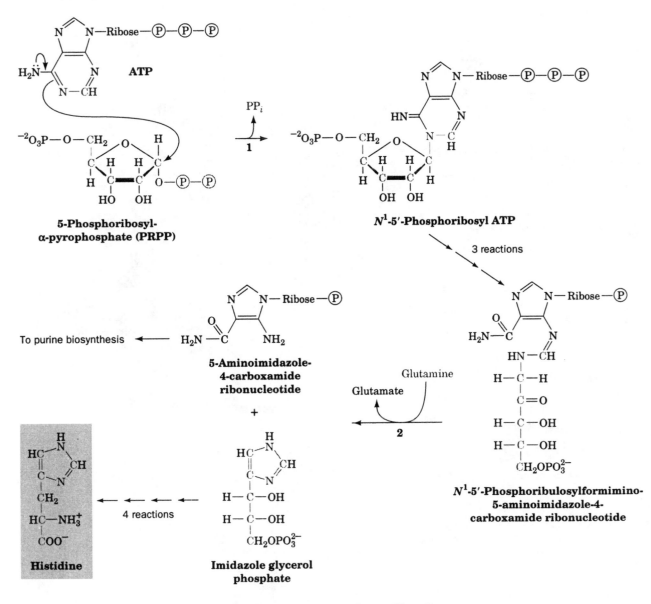

Figure 20-34. The biosynthesis of histidine.

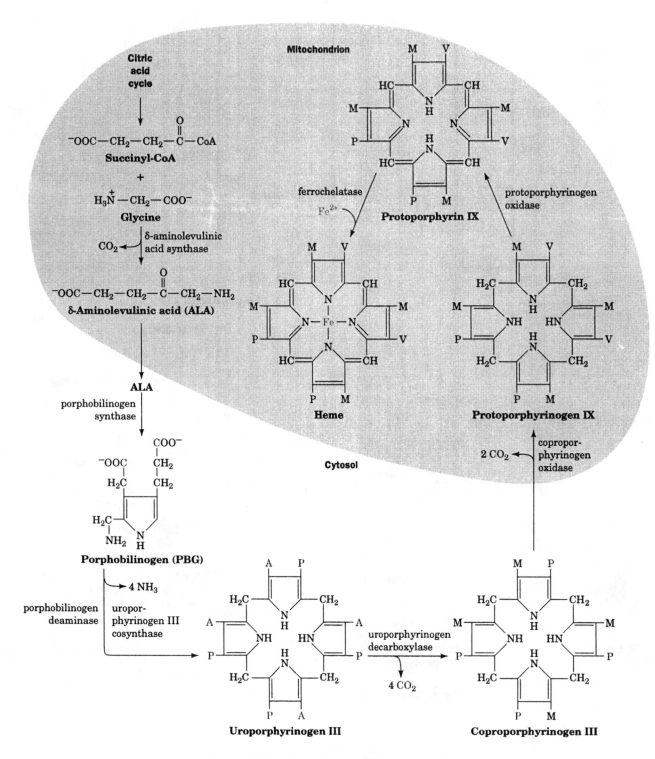

Figure 20-35. The pathway of heme biosynthesis.

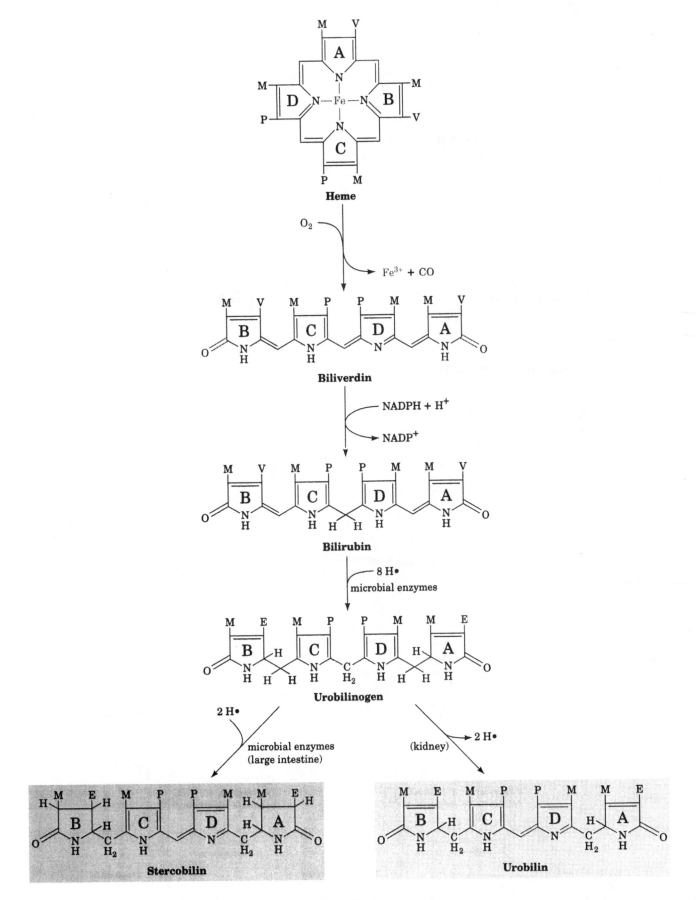

Figure 20-36. Pathway for heme degradation.

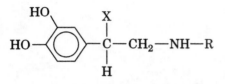

X = OH, R = CH₃   **Epinephrine (Adrenalin)**
X = OH, R = H   **Norepinephrine**
X = H,   R = H   **Dopamine**

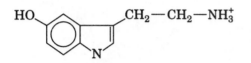

**Serotonin
(5-hydroxytryptamine)**

⁻OOC—CH₂—CH₂—CH₂—NH₃⁺

γ-Aminobutyric acid (GABA)

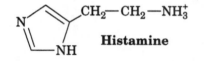

Histamine

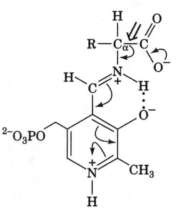

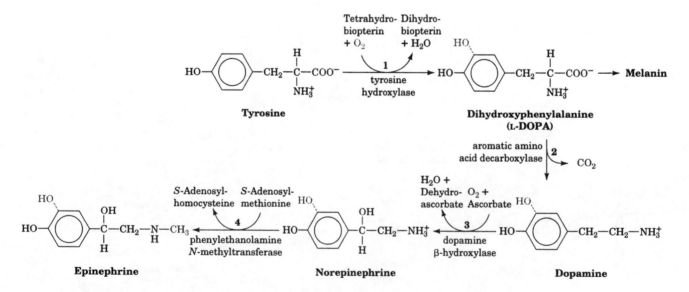

Figure 20-37. The sequential synthesis of L-DOPA, dopamine, norepinephrine, and epinephrine from tyrosine.

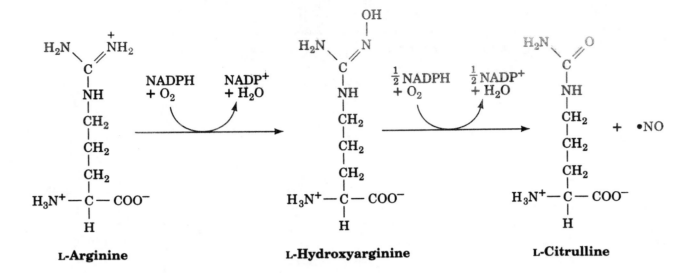

L-Arginine · L-Hydroxyarginine · L-Citrulline

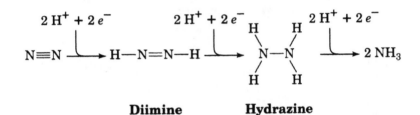

Diimine · **Hydrazine**

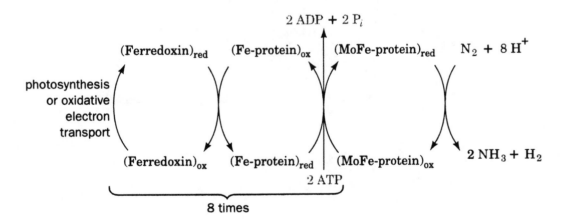

Figure 20-42. The flow of electrons in the nitrogenase-catalyzed reduction of N_2.

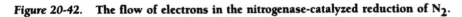

CHAPTER 21

MAMMALIAN FUEL METABOLISM: INTEGRATION AND REGULATION

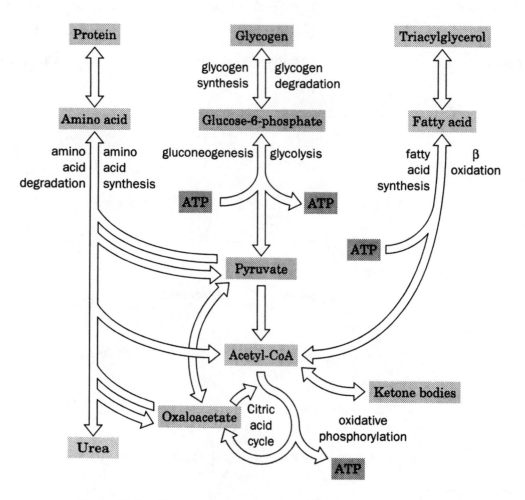

Figure 21-1. **The major pathways of fuel metabolism in mammals.**

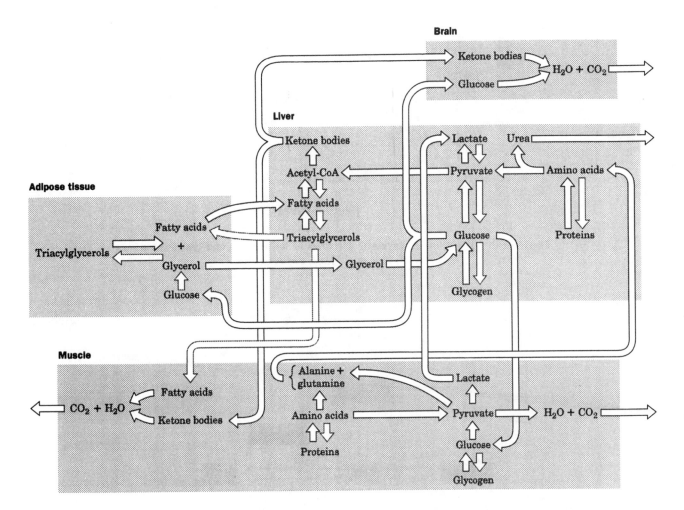

Figure 21-2. *Key to Metabolism.* **The metabolic interrelationships among brain, adipose tissue, muscle, and liver.**

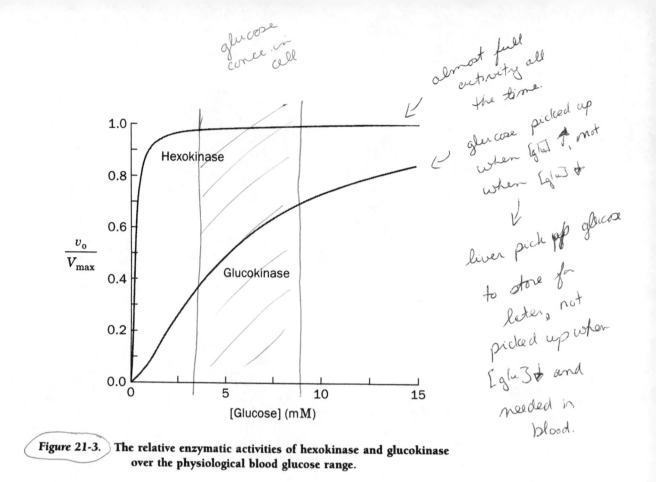

Figure 21-3. The relative enzymatic activities of hexokinase and glucokinase over the physiological blood glucose range.

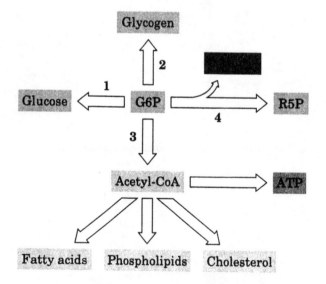

Figure 21-4. Metabolic fate of glucose-6-phosphate in liver.

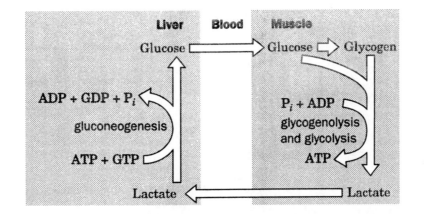

Figure 21-5. The Cori cycle.

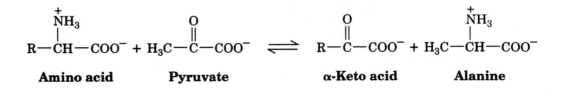

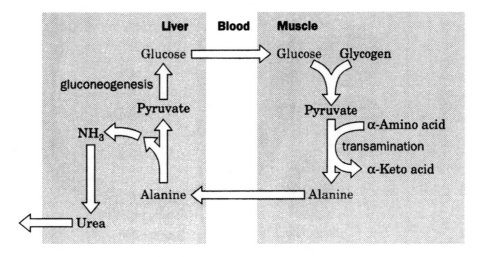

Figure 21-6. The glucose–alanine cycle.

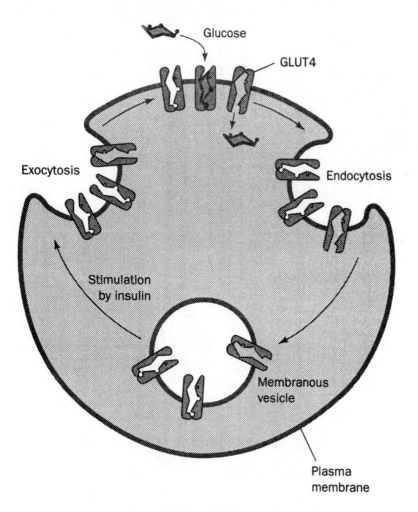

Figure 21-7.   GLUT4 activity.

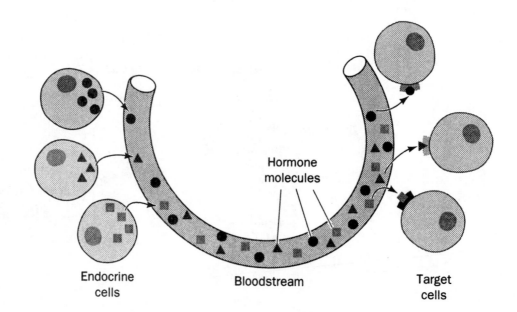

Figure 21-8.   Endocrine signaling.

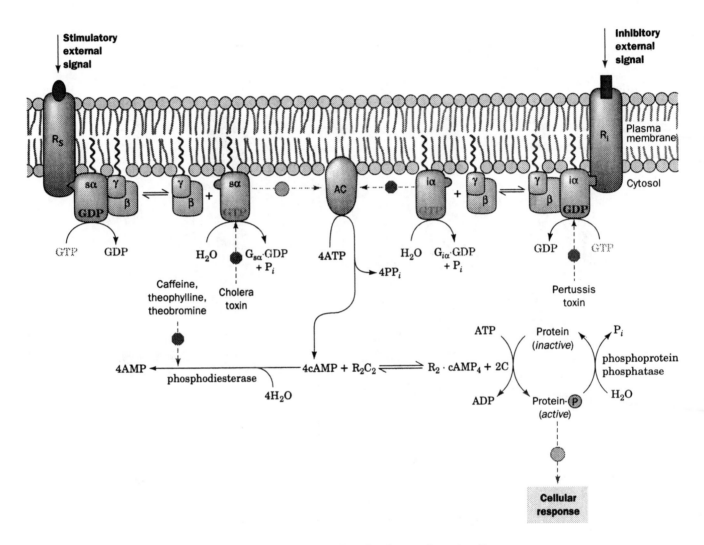

Figure 21-12. Key to Function. The adenylate cyclase signaling system.

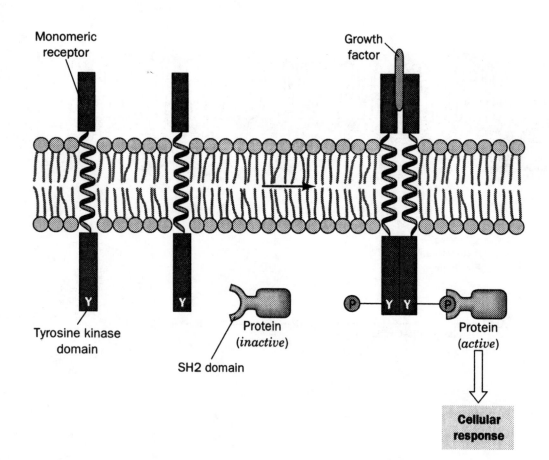

Figure 21-15. Receptor tyrosine kinase activation.

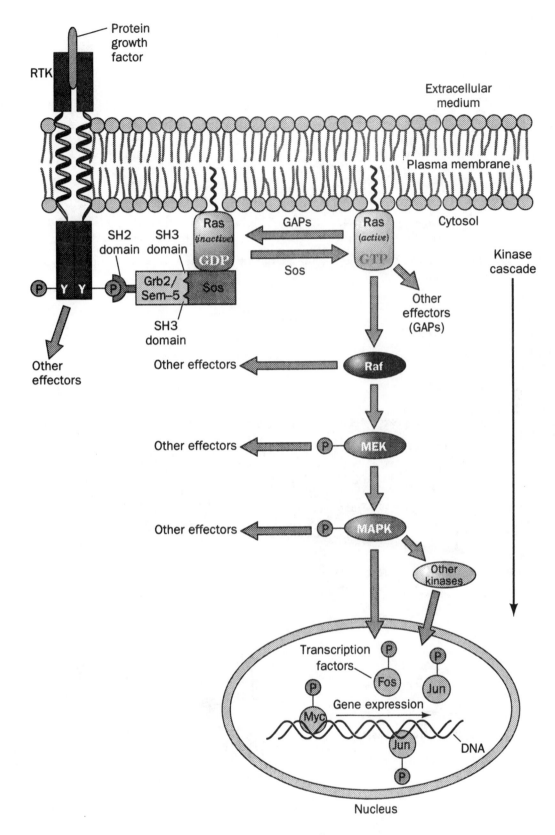

Figure 21-16. **The Ras signaling cascade.**

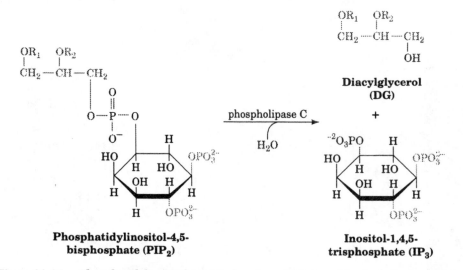

Figure 21-18. Phosphatidylinositol-4,5-bisphosphate (PIP_2) and its hydrolysis products.

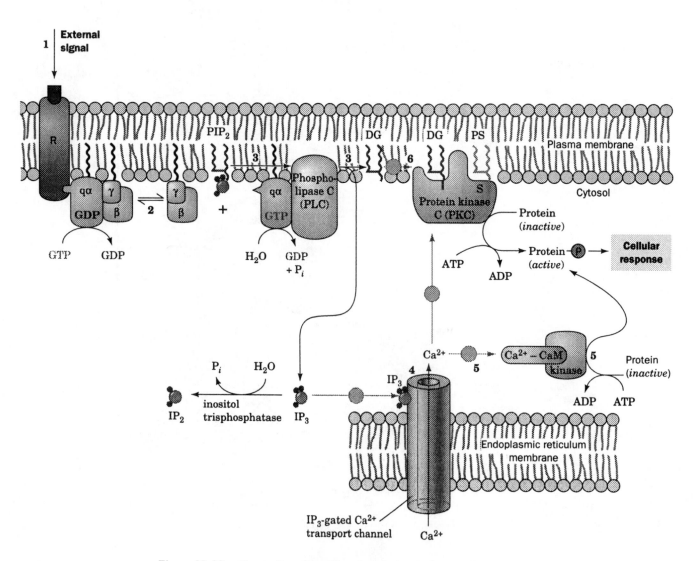

Figure 21-19. *Key to Function.* The phosphoinositide signaling system.

Table 21-1. Fuel Reserves for a Normal 70-kg Man

Fuel	Mass (kg)	Calories[a]
Tissues		
Fat (adipose triacyglycerols)	15	141,000
Protein (mainly muscle)	6	24,000
Glycogen (muscle)	0.150	600
Glycogen (liver)	0.075	300
Circulating fuels		
Glucose (extracellular fluid)	0.020	80
Free fatty acids (plasma)	0.0003	3
Triacylglycerols (plasma)	0.003	30
Total		**166,000**

[a]1 (dieter's) Calorie = 1 kcal = 4.184 kJ.
Source: Cahill, G.F., Jr., *New Engl. J. Med.* **282**, 669 (1970).

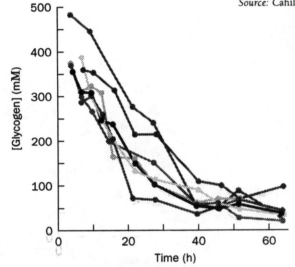

Figure 21-20. Liver glycogen depletion during fasting.

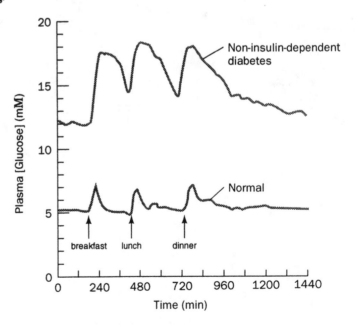

Figure 21-21. Twenty-four-hour plasma glucose profiles in normal and non-insulin-dependent diabetic subjects.

CHAPTER 22

NUCLEOTIDE METABOLISM

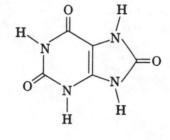

Uric acid

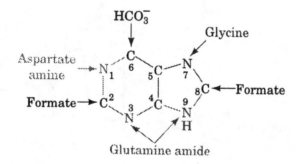

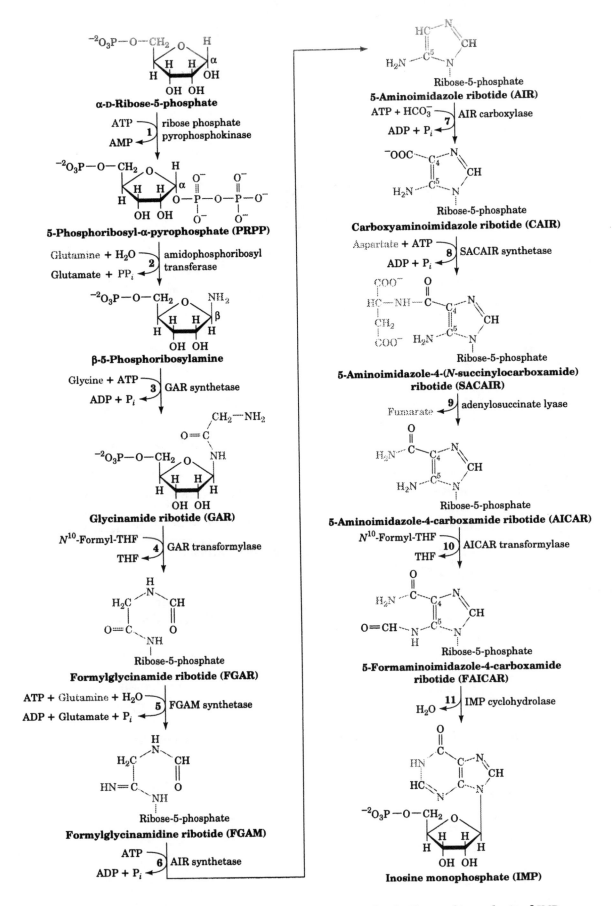

Figure 22-1. *Key to Metabolism.* The metabolic pathway for the *de novo* biosynthesis of IMP.

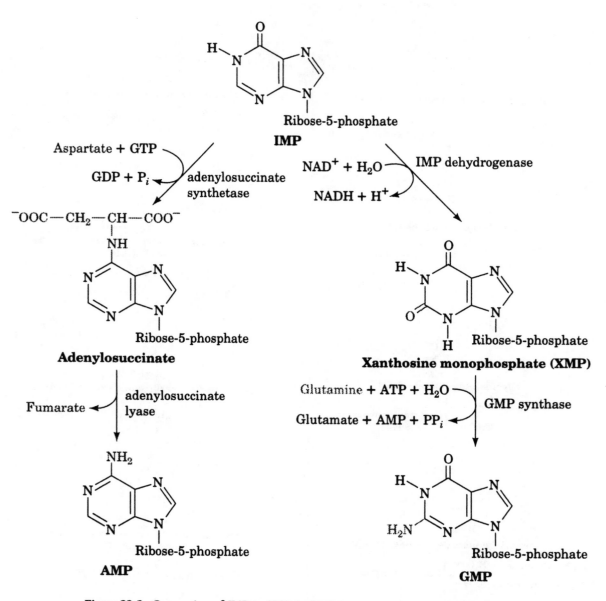

Figure 22-3. Conversion of IMP to AMP or GMP in separate two-reaction pathways.

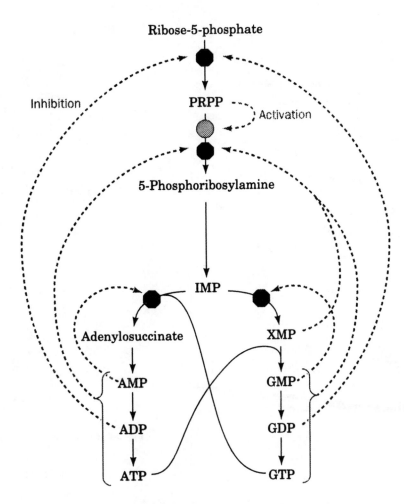

Figure 22-4. Control of the purine biosynthesis pathway.

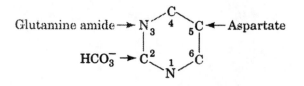

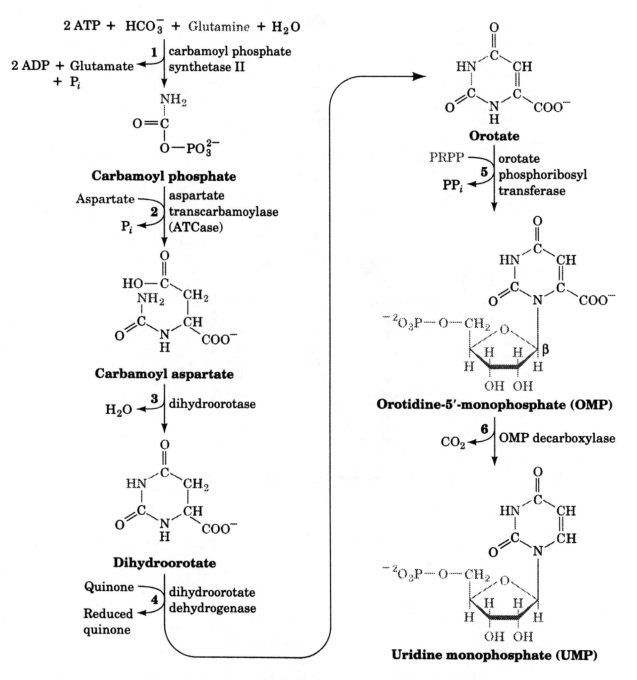

Figure 22-5.   *Key to Metabolism.* The *de novo* synthesis of UMP.

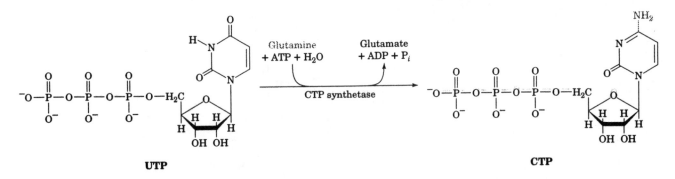

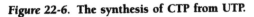

Figure 22-6. The synthesis of CTP from UTP.

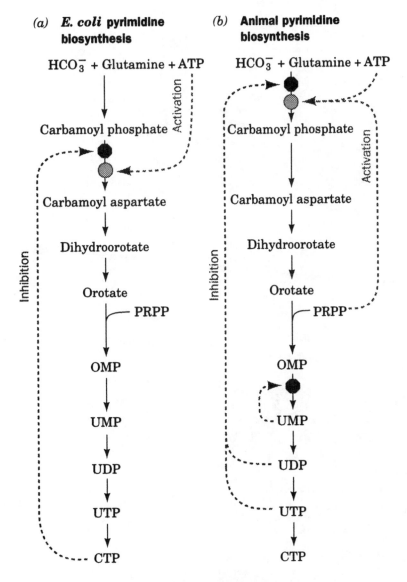

Figure 22-7. Regulation of pyrimidine biosynthesis.

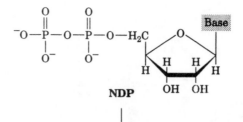

NDP

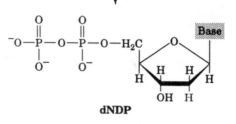

dNDP

(a)

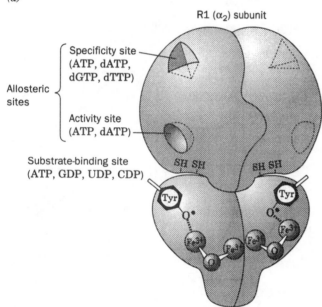

R1 (α₂) subunit

Allosteric sites {
Specificity site (ATP, dATP, dGTP, dTTP)

Activity site (ATP, dATP)
}

Substrate-binding site (ATP, GDP, UDP, CDP)

SH SH   SH SH

Tyr   Tyr

R2 (β₂) subunit

(b)   *(c)*

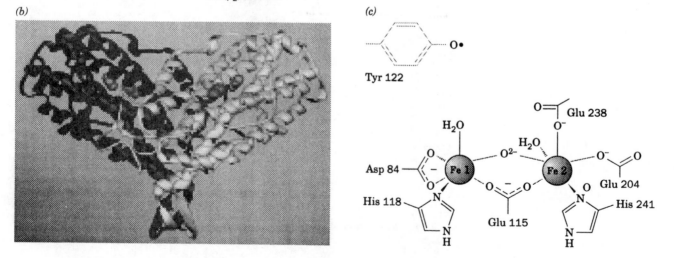

Tyr 122

Glu 238

H_2O H_2O

Asp 84   Fe 1 O^{2-} Fe 2   Glu 204

His 118   Glu 115   His 241

Figure 22-8. *E. coli* ribonucleotide reductase.

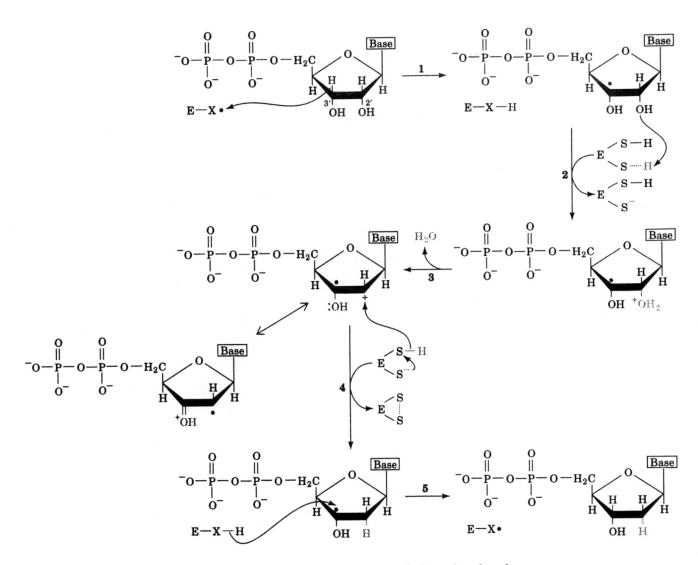

Figure 22-9. The enzymatic mechanism of ribonucleotide reductase.

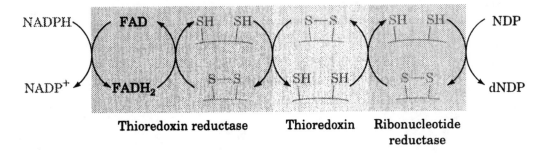

Figure 22-11. An electron-transfer pathway for nucleoside diphosphate (NDP) reduction.

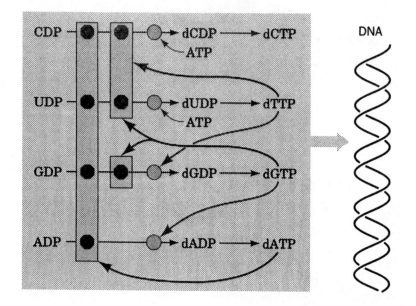

Figure 22-12. The control networks for regulating deoxyribonucleotide biosynthesis.

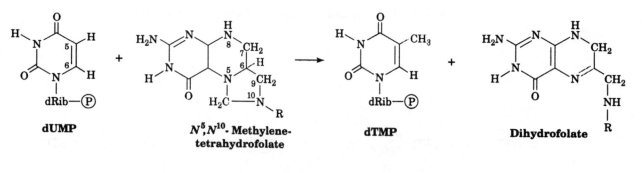

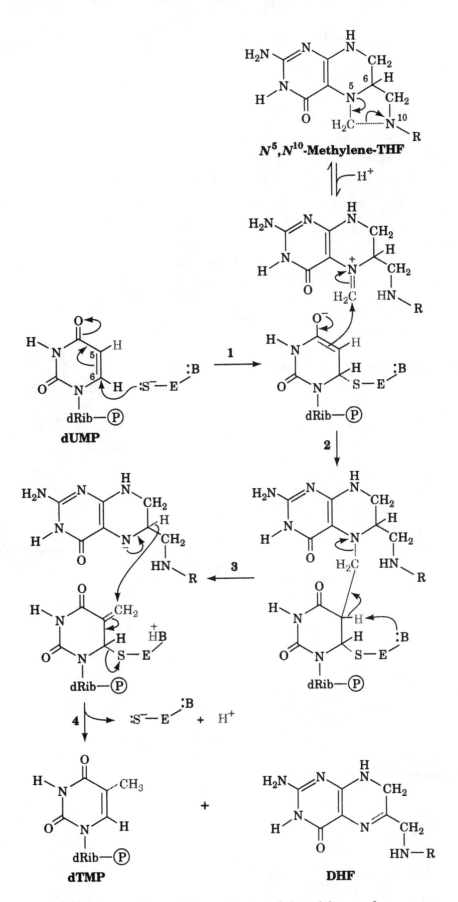

Figure 22-14. The catalytic mechanism of thymidylate synthase.

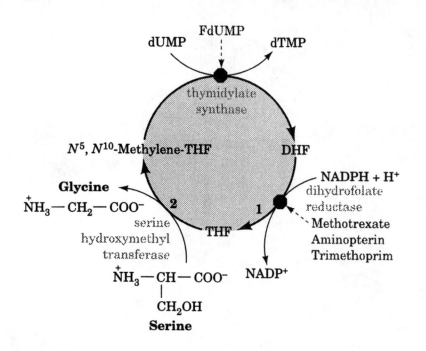

Figure 22-15. Regeneration of N^5,N^{10}-methylenetetrahydrofolate.

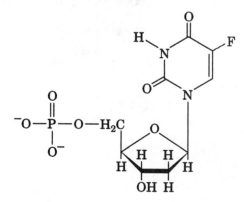

5-Fluorodeoxyuridylate (FdUMP)

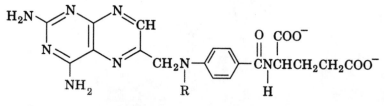

R = H   **Aminopterin**
R = CH$_3$ **Methotrexate (amethopterin)**

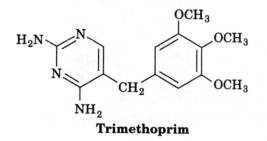

Trimethoprim

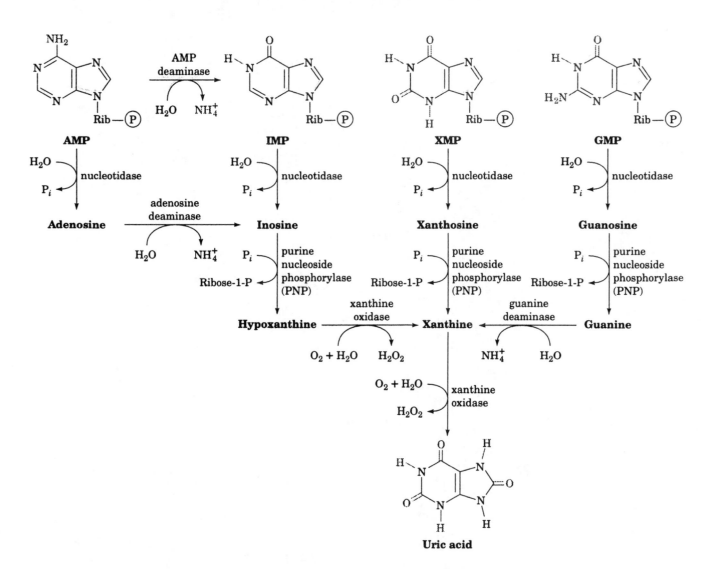

Figure 22-17. The major pathways of purine catabolism in animals.

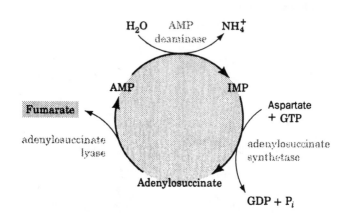

Net: H_2O + Aspartate + GTP $\longrightarrow$ NH_4^+ + GDP + P_i + fumarate

Figure 22-20. The purine nucleotide cycle.

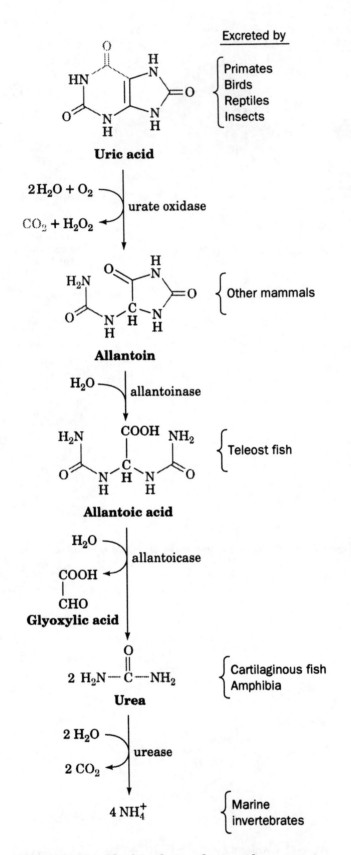

Figure 22-21. The degradation of uric acid to ammonia.

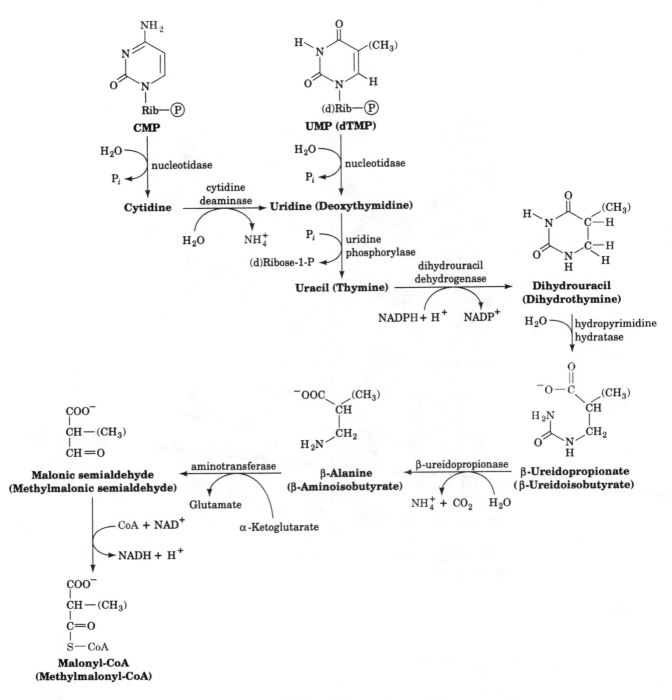

Figure 22-23. The major pathways of pyrimidine catabolism in animals.

NUCLEIC ACID STRUCTURE

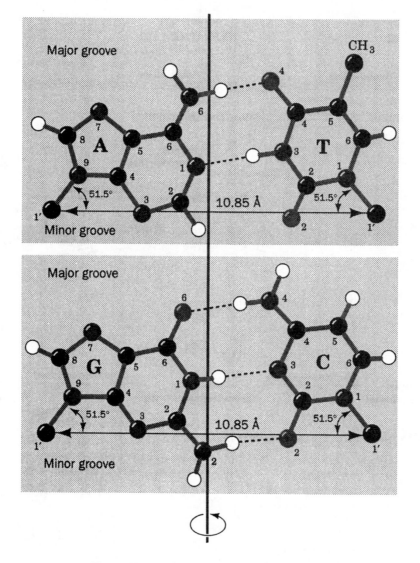

Figure 23-1. **The Watson–Crick base pairs.**

Table 23-1. *Key to Structure.* Structural Features of Ideal A-, B-, and Z-DNA

	A	B	Z
Helical sense	Right handed	Right handed	Left handed
Diameter	~26 Å	~20 Å	~18 Å
Base pairs per helical turn	11	10	12 (6 dimers)
Helical twist per base pair	33°	36°	60° (per dimer)
Helix pitch (rise per turn)	28 Å	34 Å	45 Å
Helix rise per base pair	2.6 Å	3.4 Å	3.7 Å
Base tilt normal to the helix axis	20°	6°	7°
Major groove	Narrow and deep	Wide and deep	Flat
Minor groove	Wide and shallow	Narrow and deep	Narrow and deep
Sugar pucker	C3′-*endo*	C2′-*endo*	C2′-*endo* for pyrimidines; C3′-*endo* for purines
Glycosidic bond	Anti	Anti	Anti for pyrimidines; syn for purines

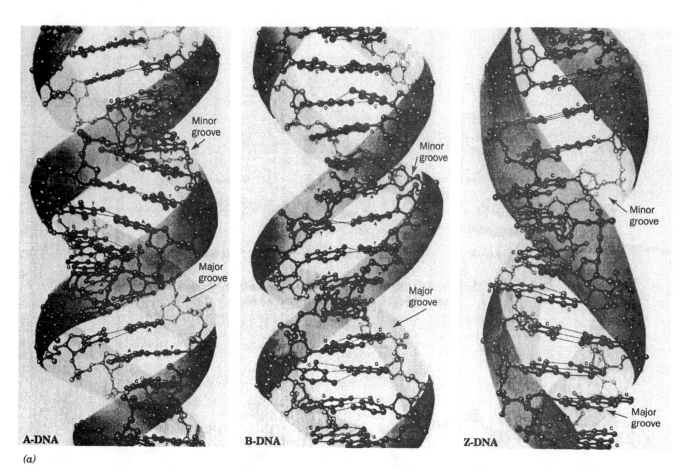

(a)

Figure 23-2. Structures of A-, B-, and Z-DNA.

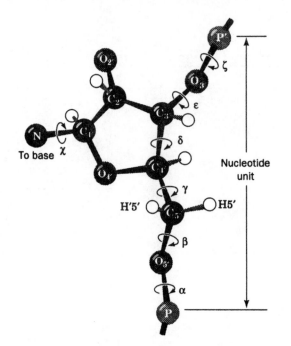

Figure 23-4. The seven torsion angles that determine the conformation of a nucleotide unit.

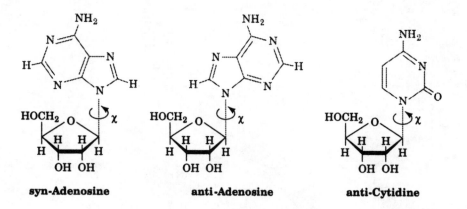

syn-Adenosine   anti-Adenosine   anti-Cytidine

Figure 23-5. The sterically allowed orientations of purine and pyrimidine bases with respect to their attached ribose units.

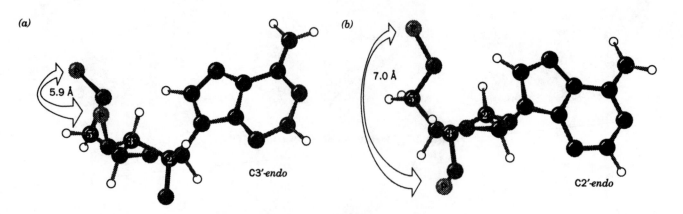

(a)   5.9 Å   C3'-endo   *(b)*   7.0 Å   C2'-endo

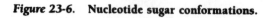

Figure 23-6. Nucleotide sugar conformations.

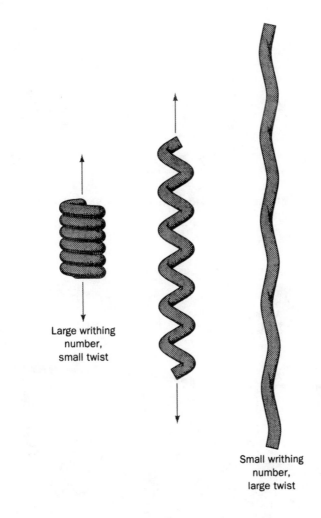

Large writing
number,
small twist

Small writhing
number,
large twist

Figure 23-8. **The difference between writhing and twist as demonstrated by a coiled telephone cord.**

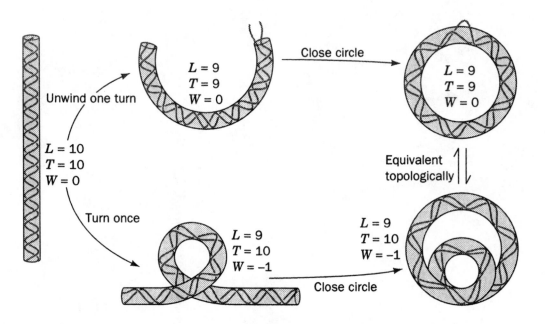

Figure 23-9. Two ways of introducing one supercoil into a DNA that has 10 duplex turns.

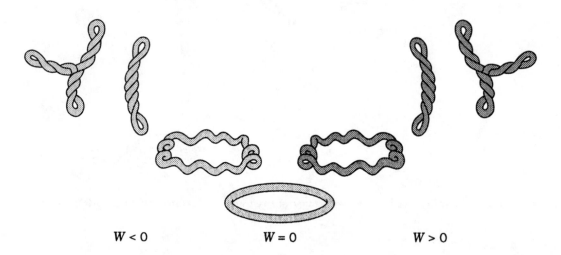

$W < 0$ $W = 0$ $W > 0$

Figure 23-10. Progressive unwinding of a negatively supercoiled DNA molecule.

320

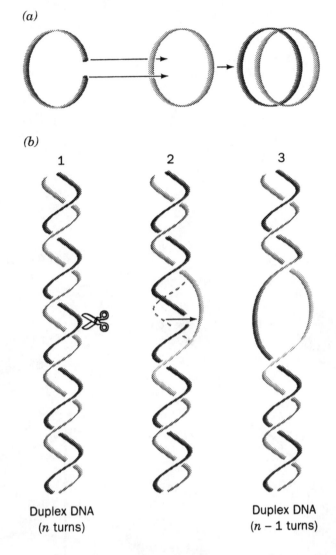

(a)

(b)

1   2   3

Duplex DNA   Duplex DNA
(n turns)   ($n-1$ turns)

Figure 23-11. **Type I topoisomerase action.**

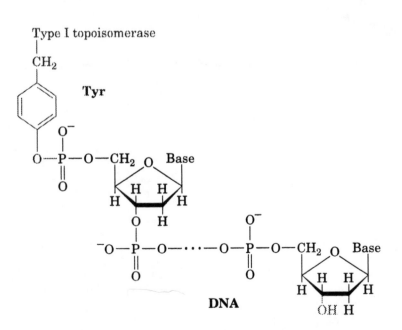

Type I topoisomerase

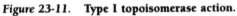

321

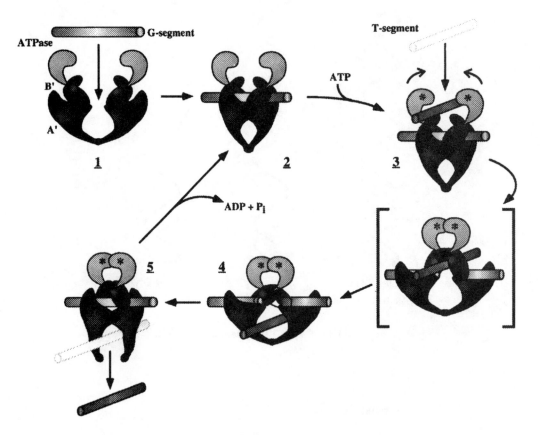

Figure 23-14. Proposed mechanism for Type II topoisomerase.

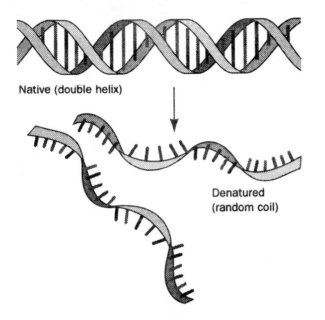

Figure 23-15. A schematic representation of DNA denaturation.

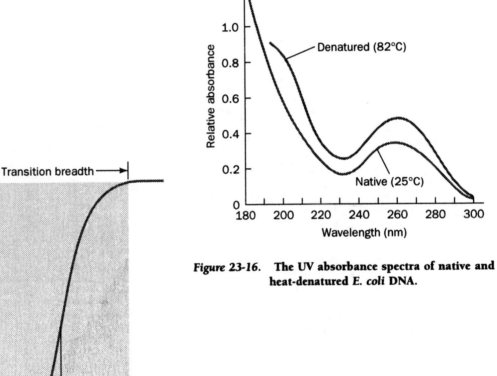

Figure 23-16. The UV absorbance spectra of native and heat-denatured *E. coli* DNA.

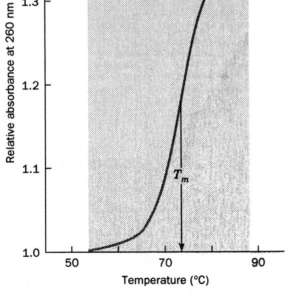

Figure 23-17. An example of a DNA melting curve.

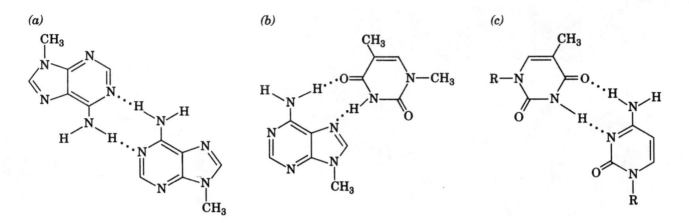

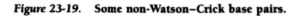

Figure 23-19. Some non-Watson–Crick base pairs.

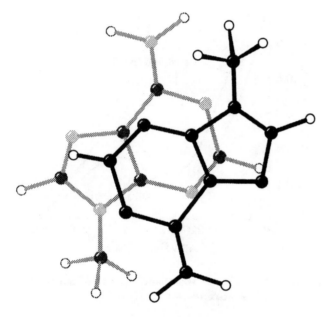

Figure 23-20. The stacking of adenine rings in the crystal structure of 9-methyladenine.

Table 23-2. Stacking Energies for the Ten Possible Dimers in B-DNA

Stacked dimer	Stacking energy $(kJ \cdot mol^{-1})$
C·G G·C	−61.0
C·G A·T	−44.0
C·G T·A	−41.0
G·C C·G	−40.5
G·C G·C	−34.6
G·C A·T	−28.4
T·A A·T	−27.5
G·C T·A	−27.5
A·T A·T	−22.5
A·T T·A	−16.0

Source: Ornstein, R.L., Rein, R., Breen, D.L., and MacElroy, R.D., *Biopolymers* **17**, 2356 (1978).

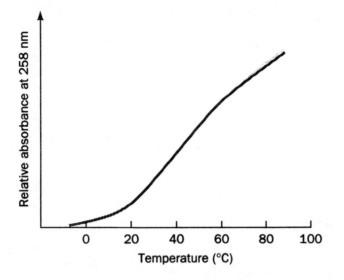

Figure 23-21. Melting curve for poly(A).

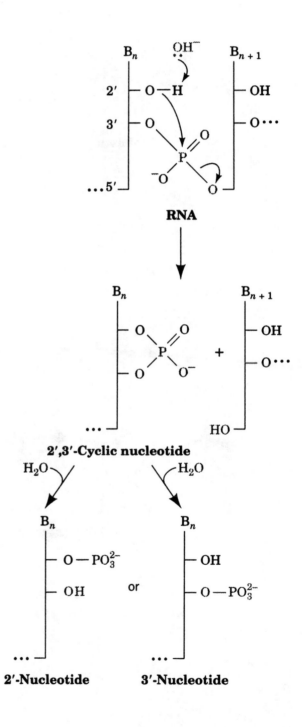

RNA

2',3'-Cyclic nucleotide

2'-Nucleotide   3'-Nucleotide

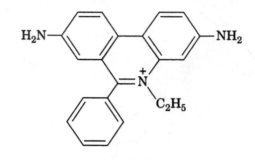

Ethidium

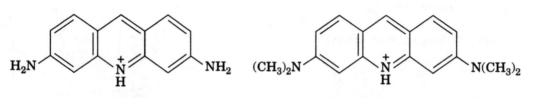

Proflavin   **Acridine orange**

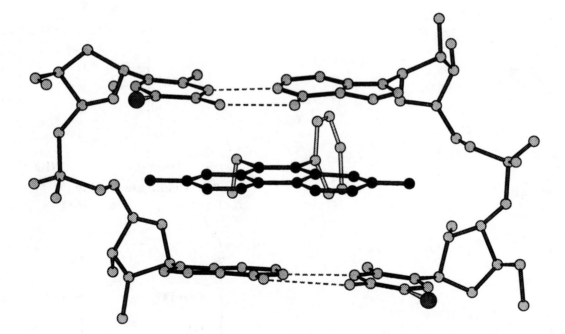

Figure 23-27. The X-ray structure of a complex of ethidium with 5-iodo-UpA.

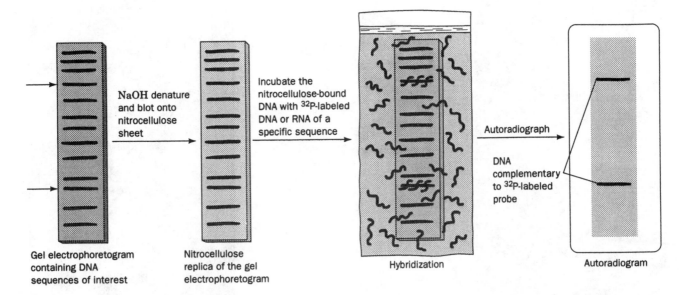

Figure 23-28. The detection of DNAs containing specific base sequences by Southern blotting.

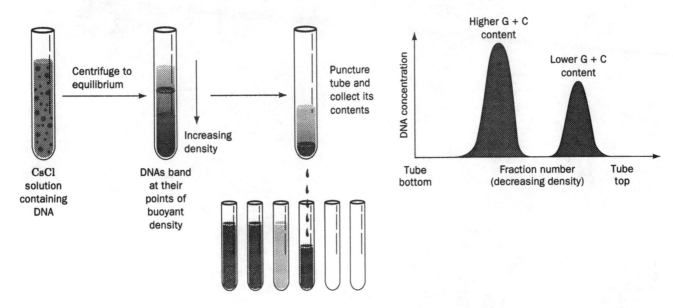

Figure 23-29. The separation of DNAs by equilibrium density gradient ultracentrifugation in CsCl solution.

(a)

(b)

(c)

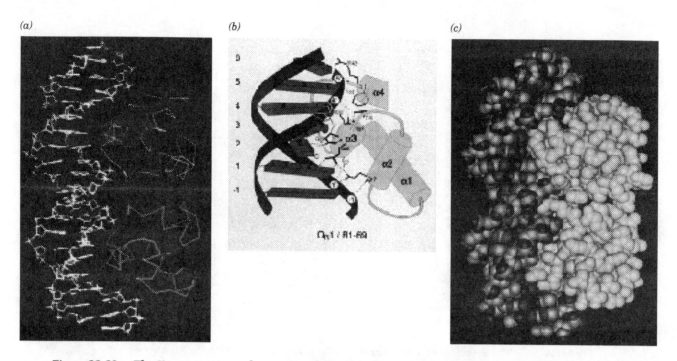

Figure 23-32. The X-ray structure of a portion of the 434 phage repressor in complex with its target DNA.

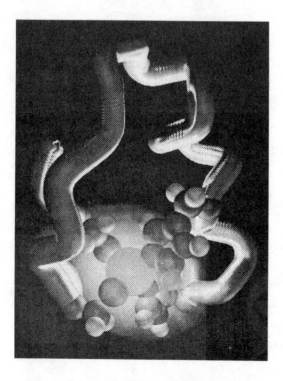

Figure 23-35. The NMR structure of a zinc finger from the *Xenopus* protein Xfin.

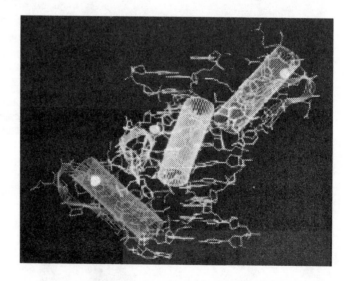

Figure 23-36. The X-ray structure of a three–zinc finger segment of Zif268 in complex with a 10-bp DNA

328

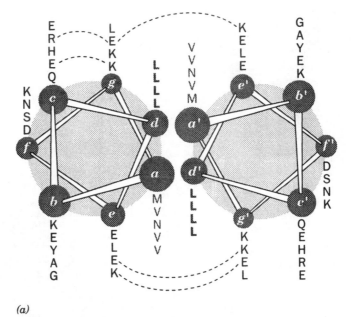

(a)

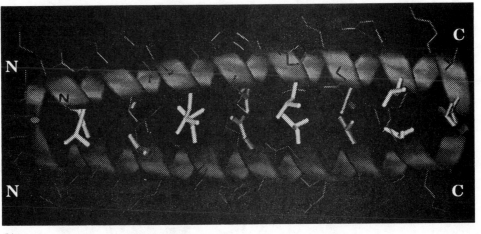

(b)

Figure 23-38. The GCN4 leucine zipper motif.

Table 23-3.   Calf Thymus Histones

Histone	Number of Residues	Mass (kD)	% Arg	% Lys
H1	215	23.0	1	29
H2A	129	14.0	9	11
H2B	125	13.8	6	16
H3	135	15.3	13	10
H4	102	11.3	14	11

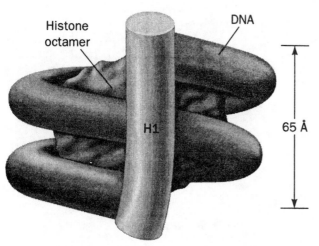

Figure 23-45.   **A model of histone H1 binding to the DNA of the 166-bp nucleosome.**

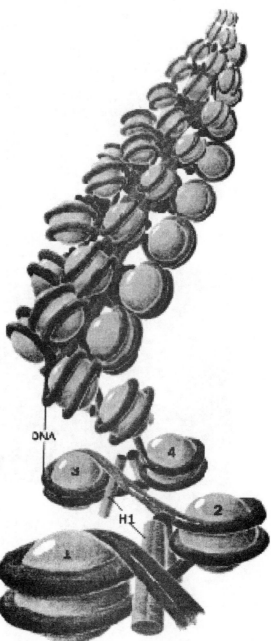

Figure 23-48.   **A proposed model of the 300-Å chromatin filament.**

CHAPTER 24

DNA REPLICATION, REPAIR, AND RECOMBINATION

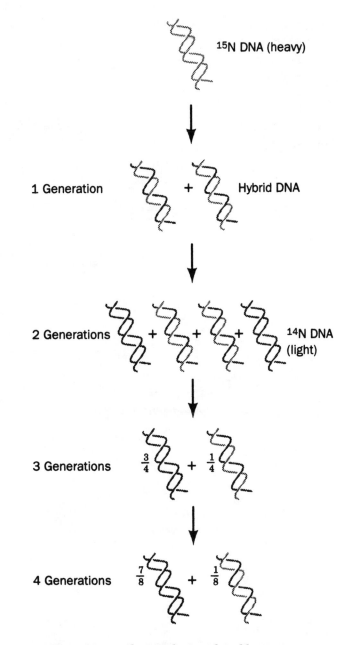

Figure 24-1. **The Meselson and Stahl experiment.**

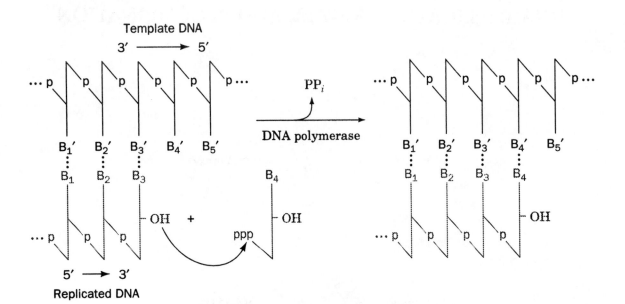

Figure 24-2. Action of DNA polymerases.

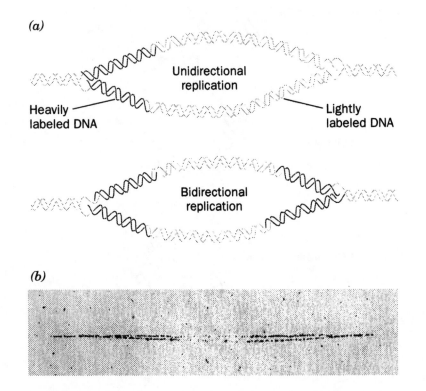

(a)

Unidirectional replication

Heavily labeled DNA

Lightly labeled DNA

Bidirectional replication

(b)

Figure 24-4. The autoradiographic differentiation of unidirectional and bidirectional θ replication of DNA.

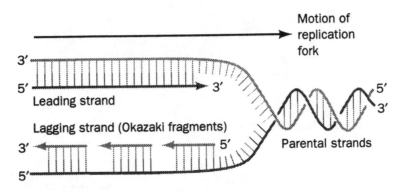

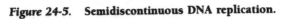

Figure 24-5. Semidiscontinuous DNA replication.

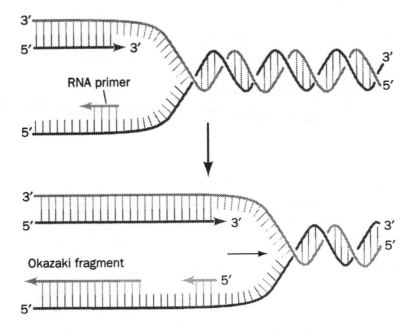

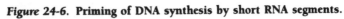

Figure 24-6. Priming of DNA synthesis by short RNA segments.

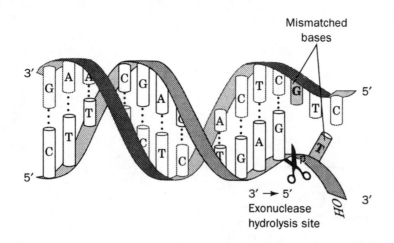

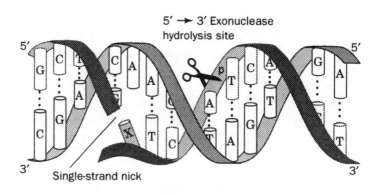

Figure 24-7. The $3' \rightarrow 5'$ exonuclease function of DNA polymerase I.

Figure 24-8. The $5' \rightarrow 3'$ exonuclease function of DNA polymerase I.

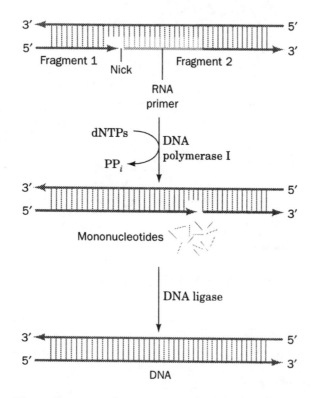

Figure 24-9. The replacement of RNA primers by DNA in lagging strand synthesis.

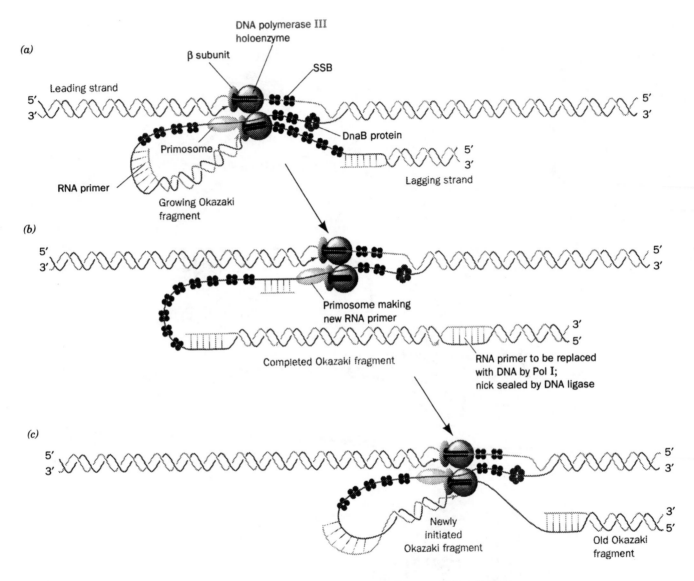

(a)

DNA polymerase III
holoenzyme

β subunit

SSB

Leading strand

5′
3′

5′
3′

DnaB protein

Primosome

RNA primer

5′
3′

Lagging strand

Growing Okazaki
fragment

(b)

5′
3′

5′
3′

Primosome making
new RNA primer

3′
5′

Completed Okazaki fragment

RNA primer to be replaced
with DNA by Pol I;
nick sealed by DNA ligase

(c)

5′
3′

5′
3′

Newly
initiated
Okazaki fragment

3′
5′

Old Okazaki
fragment

Figure 24-12. *Key to Function.* **The replication of E. coli DNA.**

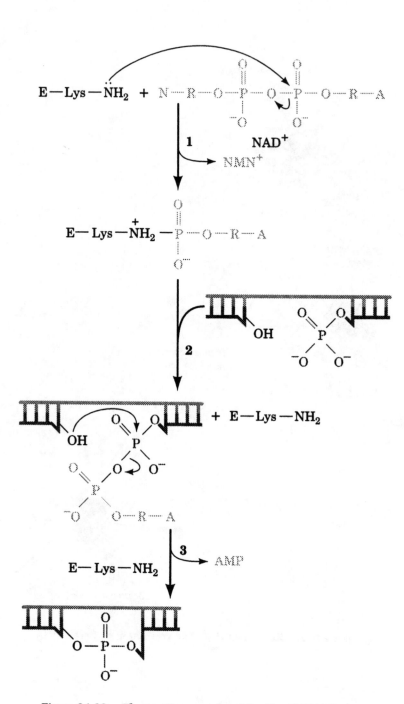

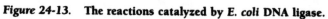

Figure 24-13. The reactions catalyzed by *E. coli* DNA ligase.

REVERSE TRANSCRIPTASE

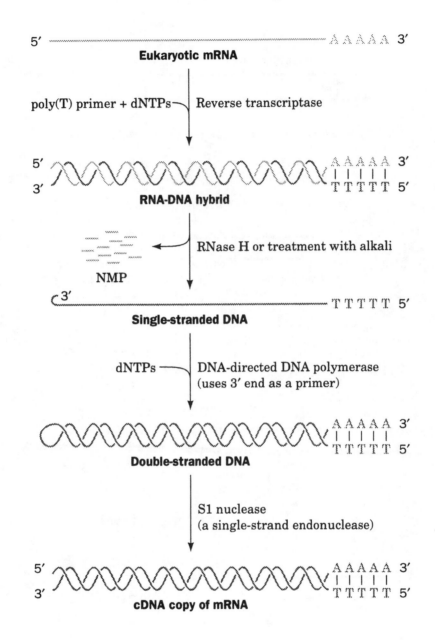

5′ ~~~~~~~~~~~~~~~~~~~~~~~~~~~~~~~~~~~~ A A A A A 3′

Eukaryotic mRNA

poly(T) primer + dNTPs ⌐ Reverse transcriptase

5′ ~~~~~~~~~~~~~~~~~~~~~~~~~~~~~~~~ A A A A A 3′
　　　　　　　　　　　　　　　　　　　 | | | | |
3′ ~~~~~~~~~~~~~~~~~~~~~~~~~~~~~~~~ T T T T T 5′

RNA-DNA hybrid

RNase H or treatment with alkali

NMP

3′ ~~~~~~~~~~~~~~~~~~~~~~~~~~~~~~~~~~~~~ T T T T T 5′

Single-stranded DNA

dNTPs ⌐ DNA-directed DNA polymerase
(uses 3′ end as a primer)

　　~~~~~~~~~~~~~~~~~~~~~~~~~~~~~~~ A A A A A 3′
　　　　　　　　　　　　　　　　　　 | | | | |
　　~~~~~~~~~~~~~~~~~~~~~~~~~~~~~~~ T T T T T 5′

Double-stranded DNA

S1 nuclease
(a single-strand endonuclease)

5′ ~~~~~~~~~~~~~~~~~~~~~~~~~~~~~~~~ A A A A A 3′
　　　　　　　　　　　　　　　　　　 | | | | |
3′ ~~~~~~~~~~~~~~~~~~~~~~~~~~~~~~~~ T T T T T 5′

cDNA copy of mRNA

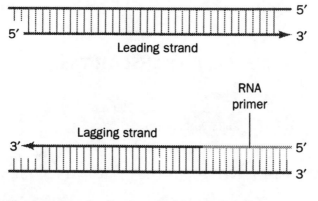

Leading strand

RNA
primer

Lagging strand

Figure 24-20. **Replication of a blunt-ended chromosome.**

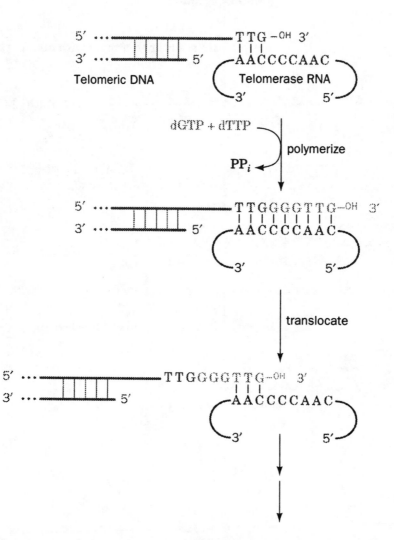

5′ ··· ————————— TTG–OH 3′
                         ｜｜｜
3′ ··· ————————— 5′ AACCCCAAC

Telomeric DNA   Telomerase RNA

                          3′   5′

dGTP + dTTP

polymerize

PPᵢ

5′ ··· ————————— TTGGGGTTG–OH 3′
                         ｜｜｜｜｜｜｜｜｜
3′ ··· ————————— 5′ AACCCCAAC

                          3′   5′

translocate

5′ ··· ————————— TTGGGGTTG–OH 3′
                              ｜｜｜｜
3′ ··· ————————— 5′   AACCCCAAC

                          3′   5′

Figure 24-21. **The proposed mechanism for the synthesis of telomeric DNA by *Tetrahymena* telomerase.**

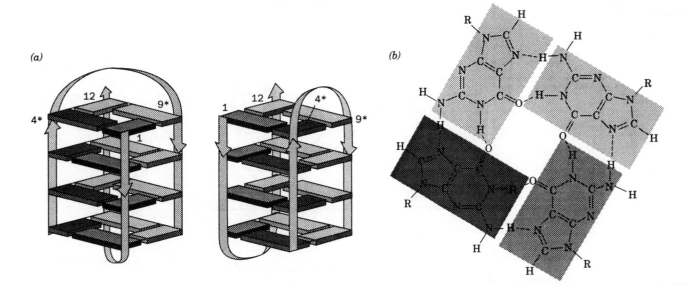

Figure 24-22. The structure of the telomeric oligonucleotide d(GGGGTTTTGGGG).

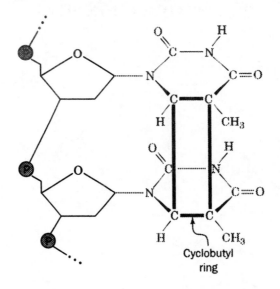

Figure 24-23. The cyclobutylthymine dimer.

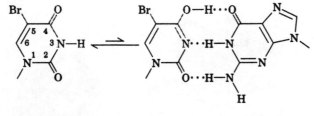

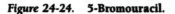

5-Bromouracil (5BU)   **5BU**
(keto tautomer)   **(enol tautomer)**   **Guanine**

Figure 24-24. 5-Bromouracil.

(a)

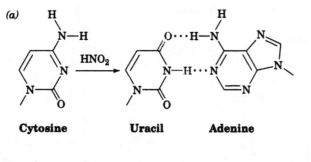

Cytosine   **Uracil**   **Adenine**

(b)

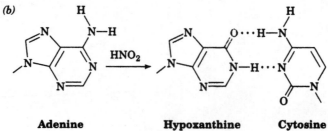

Adenine   **Hypoxanthine**   **Cytosine**

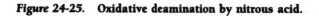

Figure 24-25. Oxidative deamination by nitrous acid.

340

DNA METHYLATION SITES

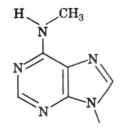

N^6-Methyladenine (m^{6}A)
residue

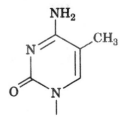

5-Methylcytosine (m^{5}C)
residue

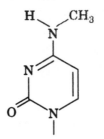

N^4-Methylcytosine (m^{4}C)
residue

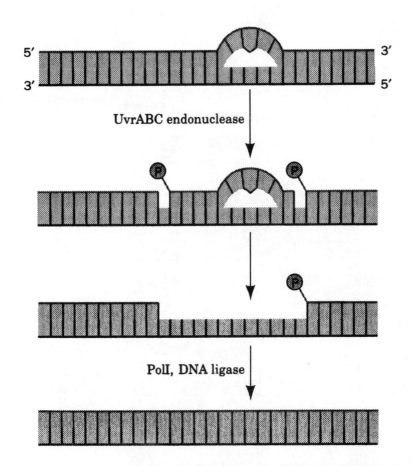

Figure 24-28. The mechanism of nucleotide excision repair (NER) of pyrimidine dimers.

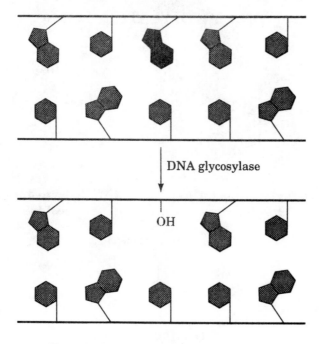

DNA glycosylase

OH

Figure 24-29. Action of DNA glycosylases.

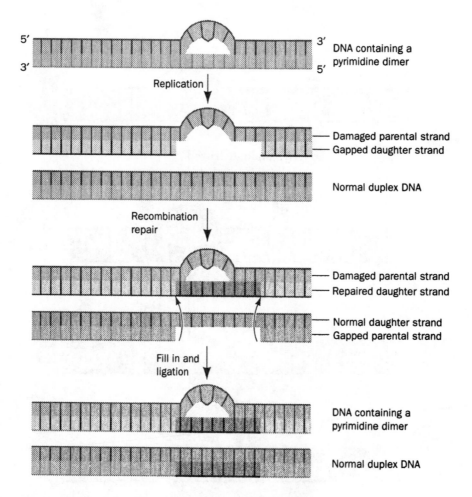

Figure 24-30. Recombination repair.

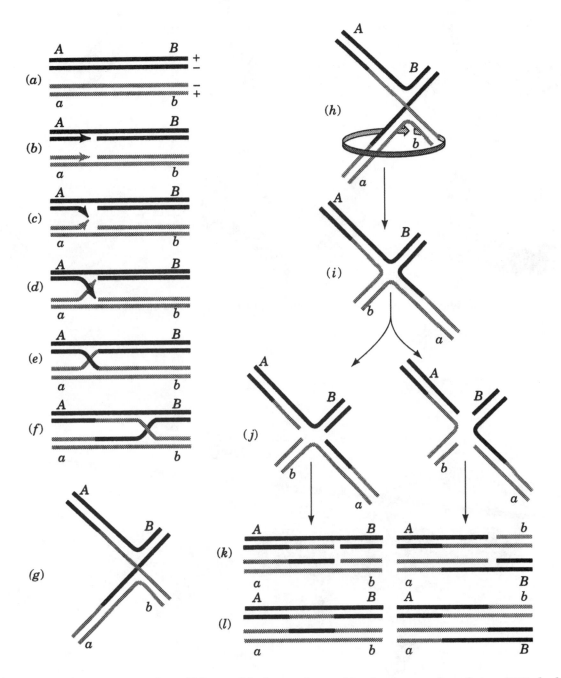

Figure 24-32. *Key to Function.* **The Holliday model of general recombination between homologous DNA duplexes.**

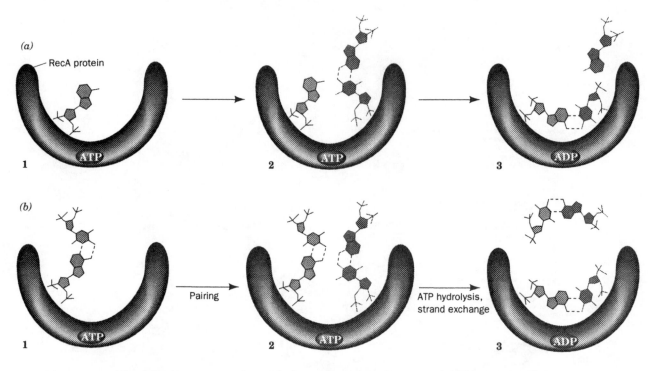

(a)

RecA protein

1   ATP   2   ATP   3   ADP

(b)

1   ATP   Pairing   2   ATP   ATP hydrolysis, strand exchange   3   ADP

Figure 24-35. Proposed models for RecA-mediated pairing and strand exchange.

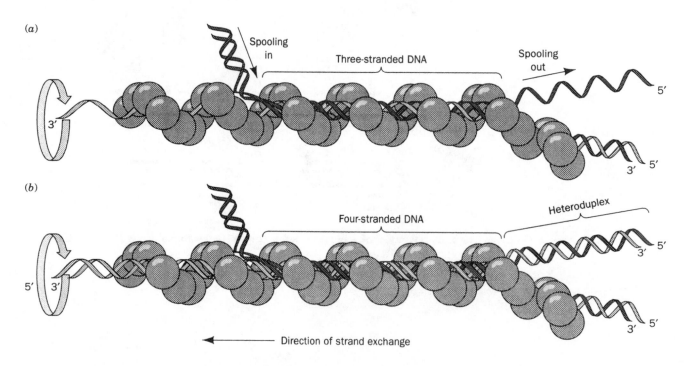

(a)

Spooling in   Three-stranded DNA   Spooling out

3′   5′

3′   5′

(b)

Four-stranded DNA   Heteroduplex

5′   3′   3′   5′

3′   5′

⟵ Direction of strand exchange

Figure 24-36. Models for three- and four-stranded DNA helices in RecA-mediated strand exchange.

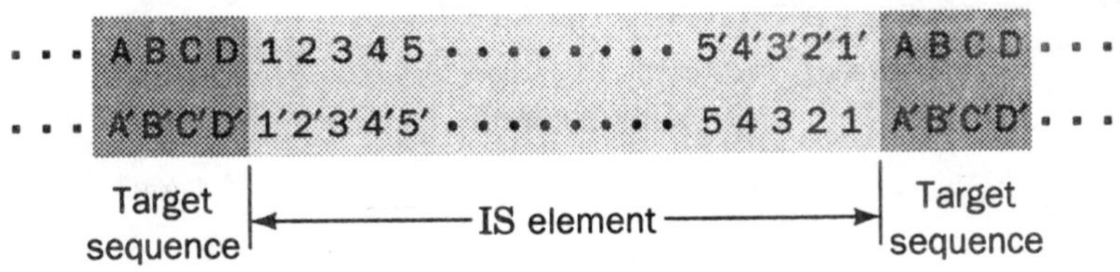

```
· · · A B C D 1 2 3 4 5 · · · · · · · · 5'4'3'2'1' A B C D · · ·
· · · A'B'C'D' 1'2'3'4'5' · · · · · · · · 5 4 3 2 1 A'B'C'D' · · ·
```

Target   ←————— IS element —————→   Target
sequence   sequence

Figure 24-38. Structure of IS elements.

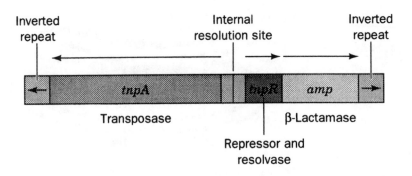

Figure 24-40. A map of transposon Tn3.

(a)

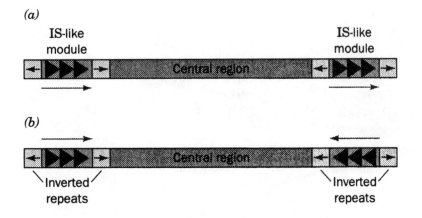

(b)

Figure 24-41. A composite transposon.

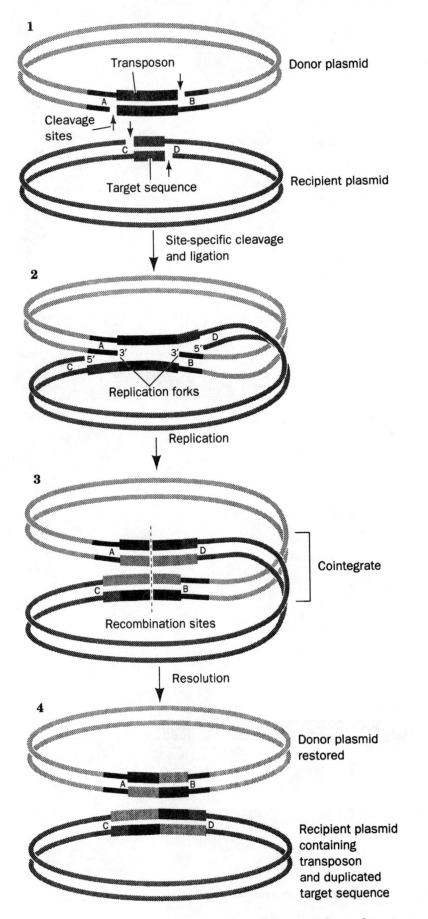

Figure 24-42. A model for transposition involving the inter-mediacy of a cointegrate.

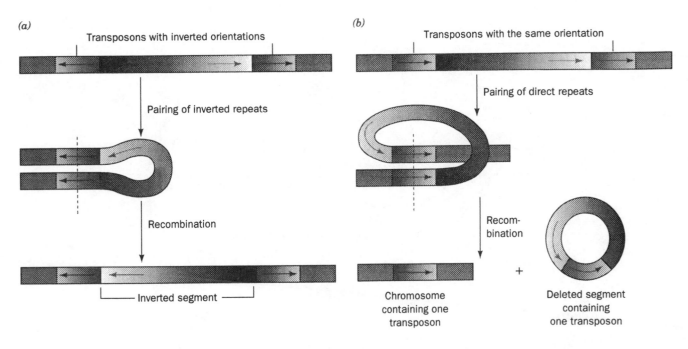

Figure 24-43. **Chromosomal rearrangement via recombination.**

TRANSCRIPTION AND RNA PROCESSING

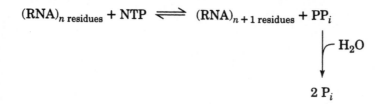

$$(RNA)_{n \text{ residues}} + NTP \rightleftharpoons (RNA)_{n+1 \text{ residues}} + PP_i$$

$$\downarrow\!\!\!-\, H_2O$$

$$2\, P_i$$

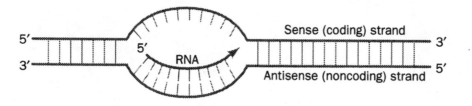

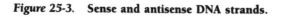

Figure 25-3.   Sense and antisense DNA strands.

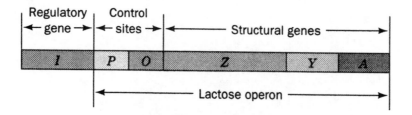

Figure 25-4.   The E. coli lac operon.

Operon	−35 region	−10 region (Pribnow box)	Initiation site (+1)

Operon	Sequence
lac	ACCCCAGGCTTTACACTTTATGCTTCCGGCTCGTATGTTGTGTGGAATTGTGAGCGG
lacI	CCATCGAATGGCGCAAAACCTTTCGCGGTATGGCATGATAGCGCCCGGAAGAGAGTC
galP2	ATTTATTCCATGTCACACTTTTCGCATCTTTGTTATGCTATGGTTATTTCATACCAT
araBAD	GGATCCTACCTGACGCTTTTTATCGCAACTCTCTACTGTTTCTCCATACCCGTTTTT
araC	GCCGTGATTATAGACACTTTTGTTACGCGTTTTTGTCATGGCTTTGGTCCCGCTTTG
trp	AAATGAGCTGTTGACAATTAATCATCGAACTAGTTAACTAGTACGCAAGTTCACGTA
bioA	TTCCAAAACGTGTTTTTTGTTGTTAATTCGGTGTAGACTTGTAAACCTAAATCTTTT
bioB	CATAATCGACTTGTAAACCAAATTGAAAAGATTTAGGTTTACAAGTCTACACCGAAT
*t*RNA^Tyr	CAACGTAACACTTTACAGCGGCGCGTCATTTGATATGATGCGCCCCGCTTCCCGATA
rrnD1	CAAAAAAATACTTGTGCAAAAAATTGGGATCCCTATAATGCGCCTCCGTTGAGACGA
rrnE1	CAATTTTTCTATTGCGGCCTGCGGAGAACTCCCTATAATGCGCCTCCATCGACACGG
rrnA1	AAAATAAATGCTTGACTCTGTAGCGGGAAGGCGTATTATGCACACCCCGCGCCGCTG

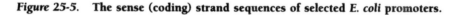

	−35 region		−10 region		Initiation site
Consensus sequence:	T T G A C A	... 16–19 bp ...	T A T A A T	... 5–8 bp ...	A 51 / C 55 / G 42 / T 48
	69 79 61 56 54 54		77 76 60 61 56 82		

Figure 25-5. The sense (coding) strand sequences of selected *E. coli* promoters.

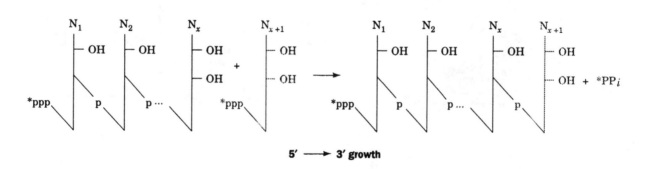

5' ⟶ 3' growth

Figure 25-6. 5' → 3' RNA chain growth.

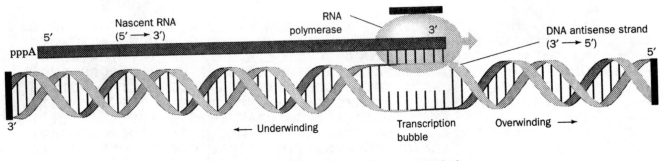

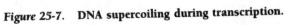

Figure 25-7. DNA supercoiling during transcription.

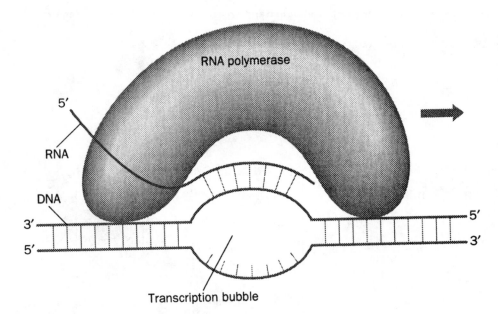

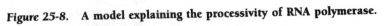

Figure 25-8. A model explaining the processivity of RNA polymerase.

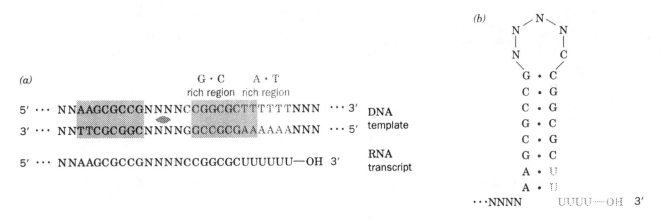

$$G \cdot C \qquad A \cdot T$$
$$\text{rich region} \quad \text{rich region}$$

5′ ··· NNAAGCGCCGNNNNCCGGCGCTTTTTTNNN ··· 3′   DNA
3′ ··· NNTTCGCGGCNNNNGGCCGCGAAAAAAANNN ··· 5′   template

5′ ··· NNAAGCGCCGNNNNCCGGCGCUUUUUU—OH 3′   RNA transcript

$$
\begin{array}{c}
N \\
N \quad N \\
N \qquad C \\
G \cdot C \\
C \cdot G \\
C \cdot G \\
G \cdot C \\
C \cdot G \\
G \cdot C \\
A \cdot U \\
A \cdot U
\end{array}
$$

···NNNN   UUUU—OH   3′

Figure 25-10.   A hypothetical strong (efficient) E. coli terminator.

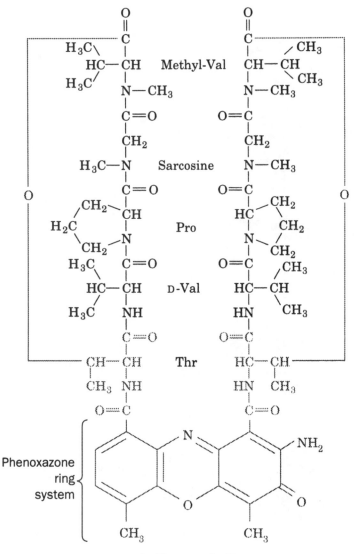

Actinomycin D

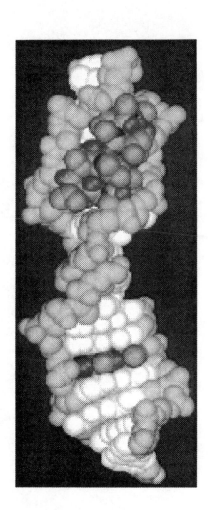

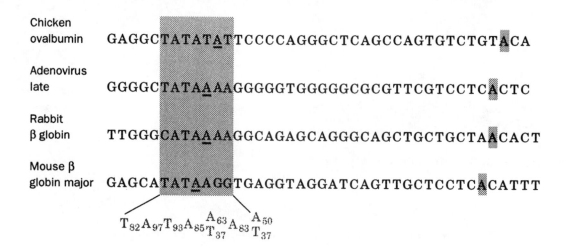

Chicken
ovalbumin   GAGGCTATATA<u>T</u>TCCCCAGGGCTCAGCCAGTGTCTGTA̲CA

Adenovirus
late   GGGGCTATA<u>A</u>AAGGGGGTGGGGGCGCGTTCGTCCTCA̲CTC

Rabbit
β globin   TTGGGCATA<u>A</u>AAGGCAGAGCAGGGCAGCTGCTGCTAA̲CACT

Mouse β
globin major   GAGCATATA<u>A</u>GGTGAGGTAGGATCAGTTGCTCCTCA̲CATTT

$$T_{82}A_{97}T_{93}A_{85}{}^{A_{63}}_{T_{37}}A_{83}{}^{A_{50}}_{T_{37}}$$

Figure 25-12. The promoter sequences of selected eukaryotic structural genes.

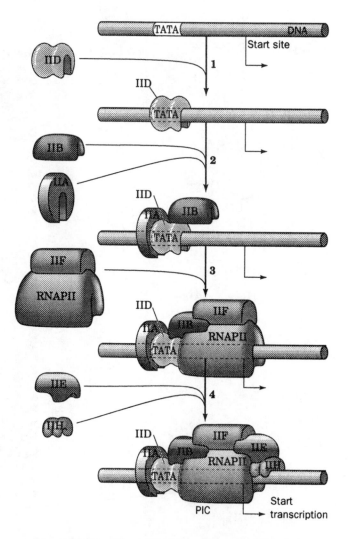

Figure 25-14. *Key to Function.* **The assembly of the preinitiation complex (PIC) on a TATA box–containing promoter.**

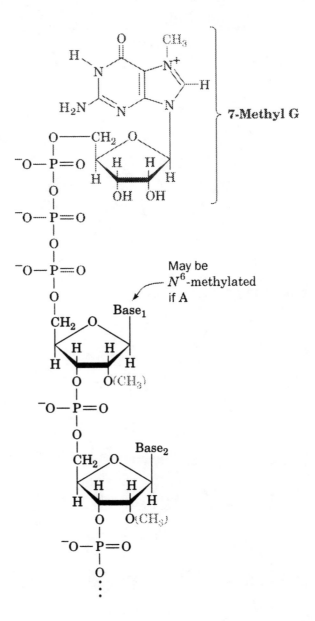

7-Methyl G

May be
N^6-methylated
if A

Figure 25-16. The structure of the 5′ cap of eukaryotic mRNAs.

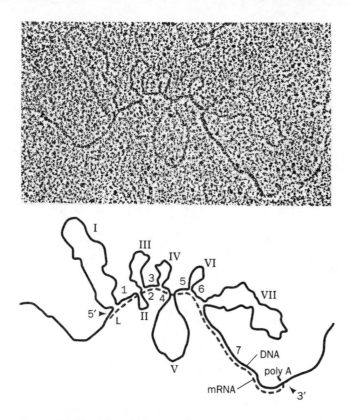

Figure 25-17. The chicken ovalbumin gene and its mRNA.

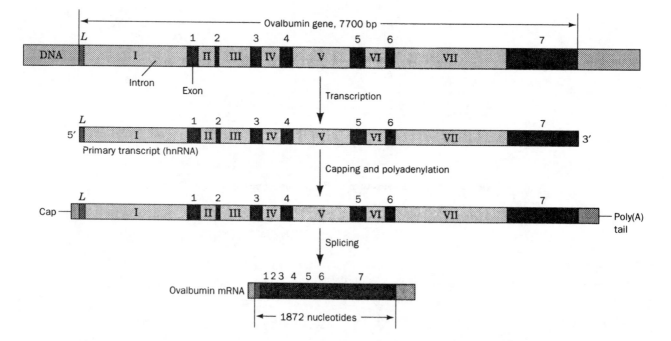

Figure 25-18. The sequence of steps in the production of mature eukaryotic mRNA.

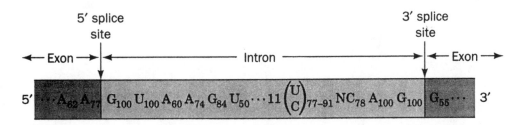

Figure 25-19. The consensus sequences at the exon–intron junctions of eukaryotic pre-mRNAs.

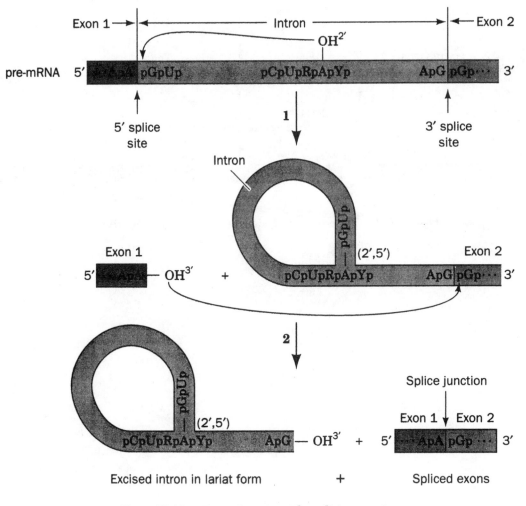

Figure 25-20. *Key to Function*. The splicing reaction.

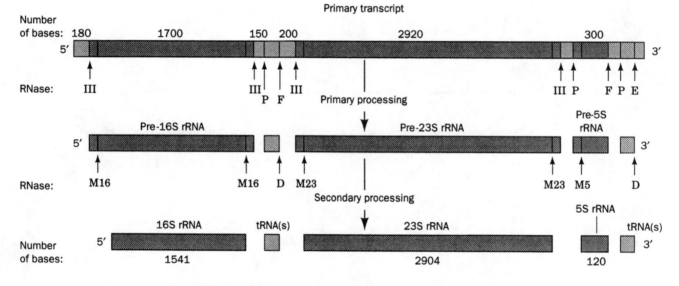

Figure 25-21. The posttranscriptional processing of *E. coli* rRNA.

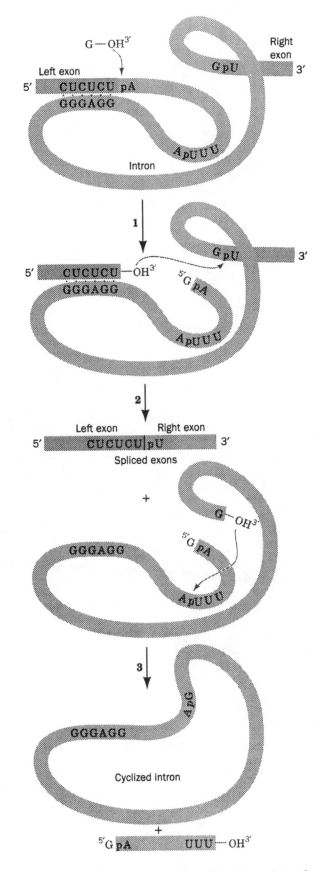

Figure 25-23. The sequence of reactions in the self-splicing of *Tetrahymena* pre-rRNA.

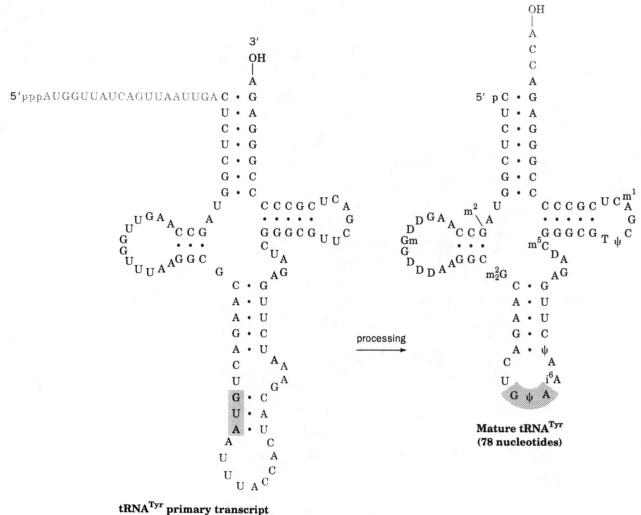

Figure 25-26. The posttranscriptional processing of yeast tRNA^{Tyr}.

CHAPTER 26

TRANSLATION

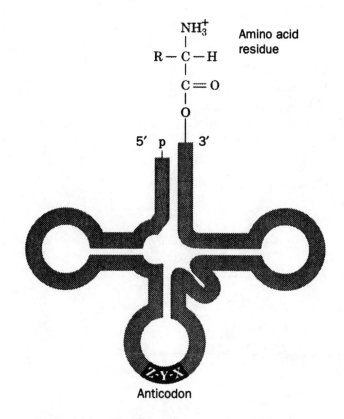

Figure 26-1. **Transfer RNA in its "cloverleaf" form.**

The structure at top (amino acid general structure):

$$H_3\overset{+}{N}-\overset{\overset{\displaystyle H}{|}}{\underset{\underset{\displaystyle R}{|}}{C}}-C\overset{\displaystyle O}{\underset{\displaystyle O^-}{\big\langle}}$$

Table 26-1. **Key to Function. The "Standard Genetic" Code**[a]

First position (5′ end)	Second position				Third position (3′ end)
	U	**C**	**A**	**G**	
U	UUU Phe UUC UUA Leu UUG	UCU UCC Ser UCA UCG	UAU Tyr UAC UAA STOP UAG	UGU Cys UGC UGA STOP UGG Trp	U C A G
C	CUU CUC CUA Leu CUG	CCU CCC Pro CCA CCG	CAU His CAC CAA Gln CAG	CGU CGC Arg CGA CGG	U C A G
A	AUU Ile AUC AUA AUG Met[b]	ACU ACC Thr ACA ACG	AAU Asn AAC AAA Lys AAG	AGU Ser AGC AGA Arg AGG	U C A G
G	GUU GUC Val GUA GUG	GCU GCC Ala GCA GCG	GAU Asp GAC GAA Glu GAG	GGU GGC Gly GGA GGG	U C A G

[a]Nonpolar amino acid residues are tan, basic residues are blue, acidic residues are red, and polar uncharged residues are purple.
[b]AUG forms part of the initiation signal as well as coding for internal Met residues.

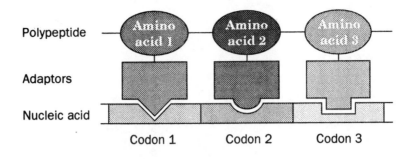

Figure 26-3. **The adaptor hypothesis.**

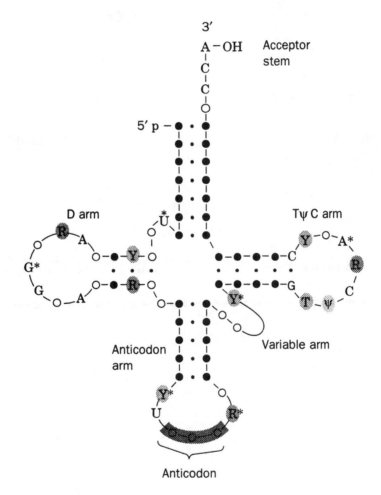

Figure 26-4. **The cloverleaf secondary structure of tRNA.**

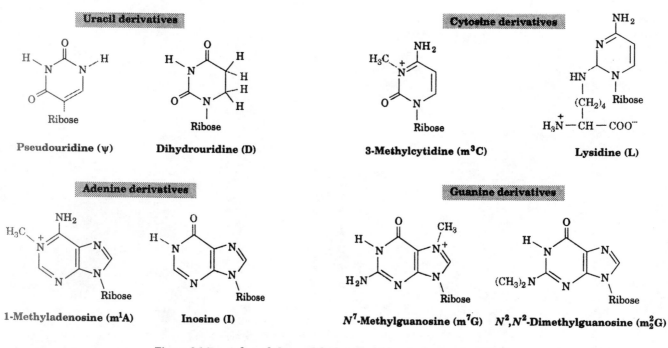

Figure 26-5. A few of the modified nucleosides that occur in tRNAs.

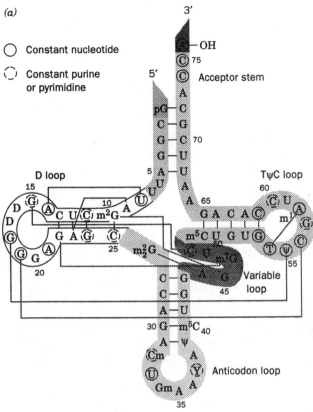

(a)

○ Constant nucleotide

◌ Constant purine or pyrimidine

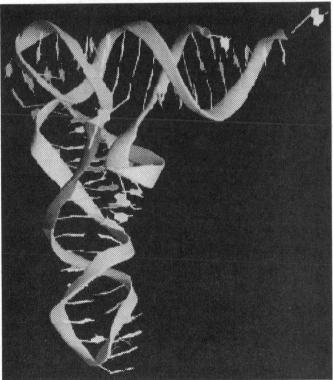

Figure 26-6. *Key to Structure.* The structure of yeast tRNA^Phe.

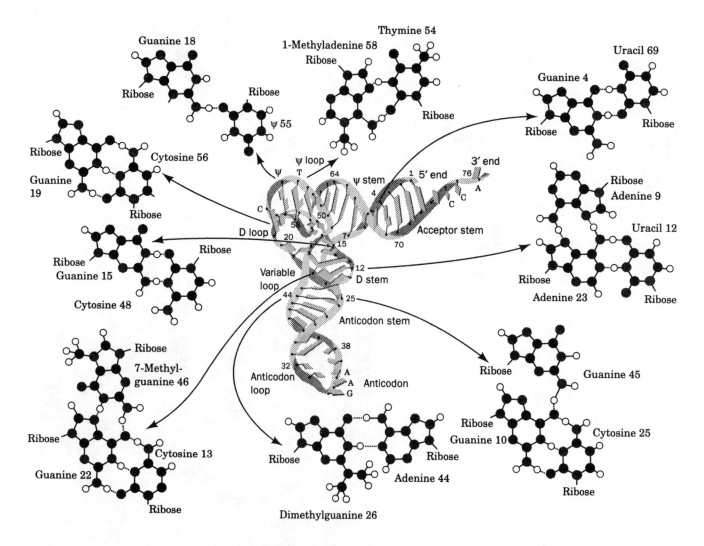

Figure 26-7. **The nine tertiary base pairing interactions in yeast tRNA^{Phe}.**

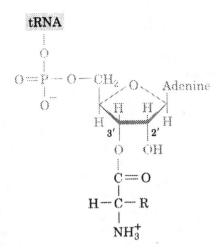

Aminoacyl–tRNA

Figure 26-8. An aminoacyl–tRNA.

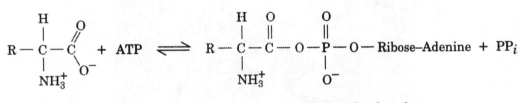

Amino acid   **Aminoacyl–adenylate**
                   (aminoacyl–AMP)

Aminoacyl–AMP + tRNA $\rightleftharpoons$ aminoacyl–tRNA + AMP

Amino acid + tRNA + ATP $\longrightarrow$ aminoacyl–tRNA + AMP + PP$_i$

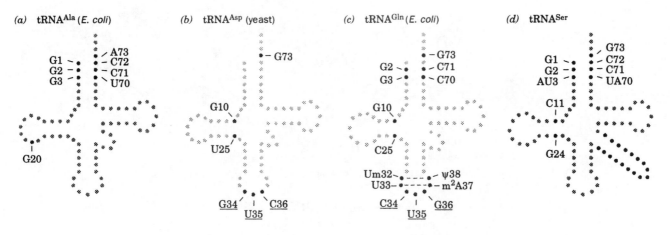

(a) tRNA^Ala (E. coli)   (b) tRNA^Asp (yeast)   (c) tRNA^Gln (E. coli)   (d) tRNA^Ser

Figure 26-9. Major identity elements in four tRNAs.

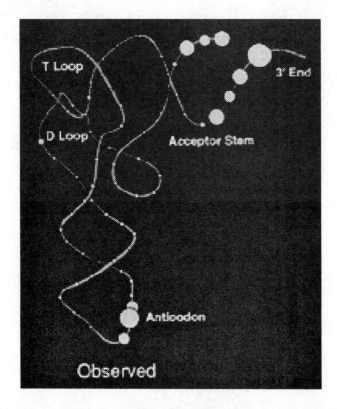

Figure 26-10. The experimentally observed identity elements of tRNAs.

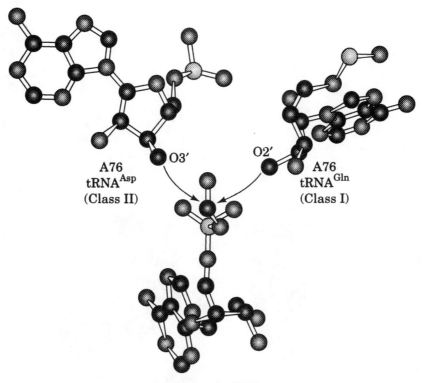

A76
tRNA^{Asp}
(Class II)

O3'

O2'

A76
tRNA^{Gln}
(Class I)

Aminoacyl–AMP

Figure 26-14. **A comparison of the stereochemistries of aminoacylation by Class I and Class II synthetases.**

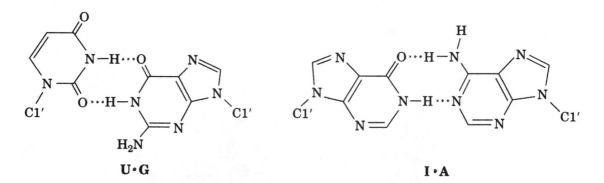

U·G

I·A

Figure 26-15. **U·G and I·A wobble pairs.**

Table 26-4. Components of *E. coli* Ribosomes

	Ribosome	Small Subunit	Large Subunit
Sedimentation coefficient	70S	30S	50S
Mass (kD)	2520	930	1590
RNA			
Major		16S, 1542 nucleotides	23S, 2904 nucleotides
Minor			5S, 120 nucleotides
RNA mass (kD)	1664	560	1104
Proportion of mass	66%	60%	70%
Proteins		21 polypeptides	31 polypeptides
Protein mass (kD)	857	370	487
Proportion of mass	34%	40%	30%

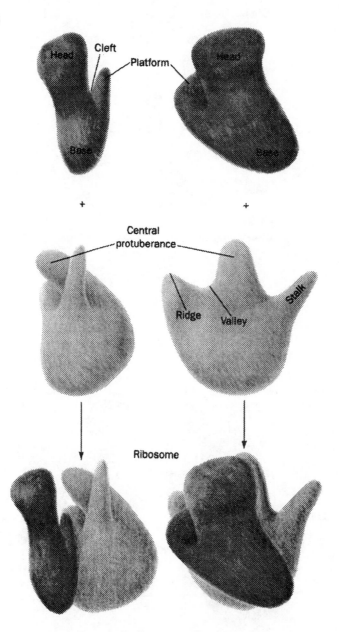

Figure 26-16. The *E. coli* ribosome.

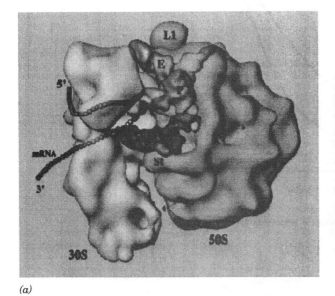
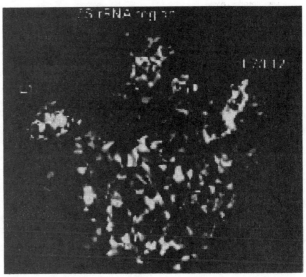

(a)   (b)

Figure 26-17.   Structure of the ribosome.

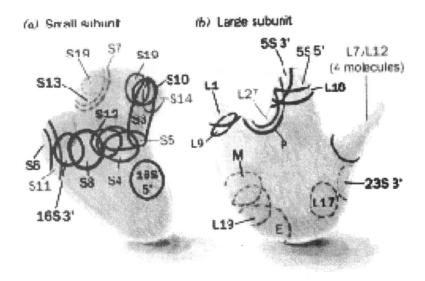

Figure 26-19.   Maps of the *E. coli* ribosomal subunits.

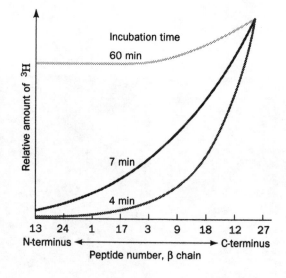

Figure 26-22. Demonstration that polypeptide synthesis proceeds from the N- to the C-terminus.

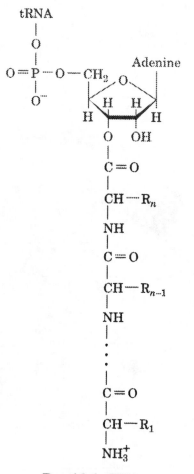

Peptidyl–tRNA

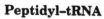

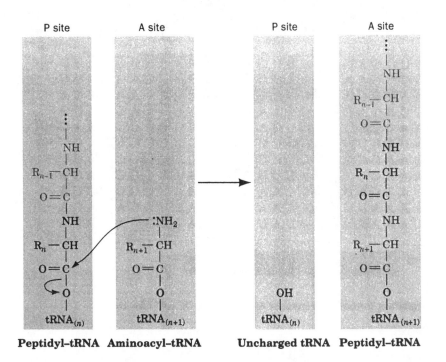

Figure 26-23. The ribosomal peptidyl transferase reaction forming a peptide bond.

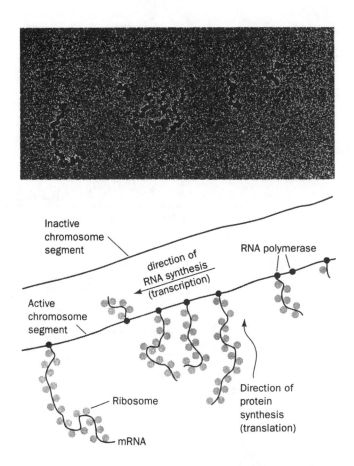

Figure 26-24. The simultaneous transcription and translation of an *E. coli* gene.

```
araB                        - U U U G G A U G G A G U G A A A C G A U G G C G A U U -
galE                        - A G C C U A A U G G A G C G A A U U A U G A G A G U U -
lacI                        - C A A U U C A G G G U G G U G A U U G U G A A A C C A -
lacZ                        - U U C A C A C A G G A A A C A G C U A U G A C C A U G -
Q β phage replicase         - U A A C U A A G G A U G A A A U G C A U G U C U A A G -
φX174 phage A protein       - A A U C U U G G A G G C U U U U U U A U G G U U C G U -
R17 phage coat protein      - U C A A C C G G G G U U U G A A G C A U G G C U U C U -
Ribosomal S12               - A A A A C C A G G A G C U A U U U A A U G G C A A C A -
Ribosomal L10               - C U A C C A G G A G C A A A G C U A A U G G C U U U A -
trpE                        - C A A A A U U A G A G A A U A A C A A U G C A A A C A -
trp leader                  - G U A A A A A G G G U A U C G A C A A U G A A A G C A -
```

```
3' end of 16S rRNA          3'  HO A U U C C U C C A C U A G -  5'
```

Figure 26-26. Some translation initiation sequences recognized by *E. coli* ribosomes.

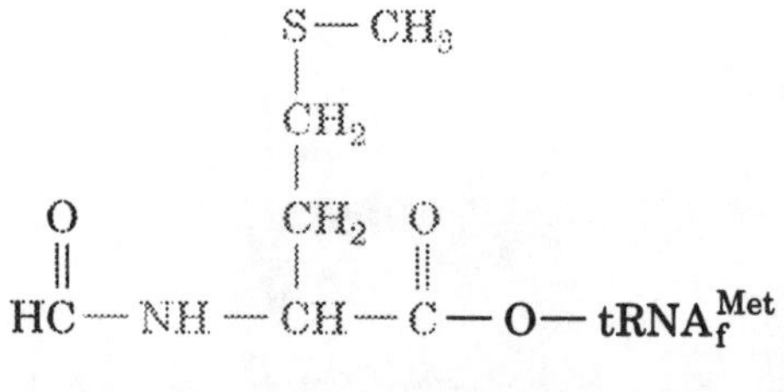

N-Formylmethionine–tRNA$_f^{Met}$
(fMet–tRNA$_f^{Met}$)

Table 26-6. The Soluble Protein Factors of *E. coli* Protein Synthesis

Factor	Mass (kD)	Function
Initiation Factors		
IF-1	9	Assists IF-3 binding
IF-2	97	Binds initiator tRNA and GTP
IF-3	22	Releases 30S subunit from inactive ribosome and aids mRNA binding
Elongation Factors		
EF-Tu	43	Binds aminoacyl–tRNA and GTP
EF-Ts	74	Displaces GDP from EF-Tu
EF-G	77	Promotes translocation by binding GTP to the ribosome
Release Factors		
RF-1	36	Recognizes UAA and UAG Stop codons
RF-2	38	Recognizes UAA and UGA Stop codons
RF-3	46	Binds GTP and stimulates RF-1 and RF-2 binding

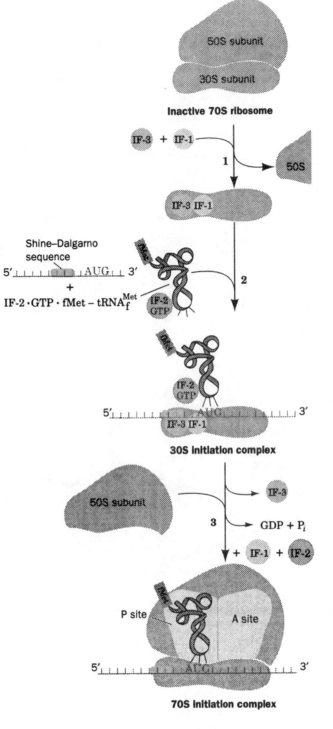

Figure 26-27. The translation initiation pathway in *E. coli* ribosomes.

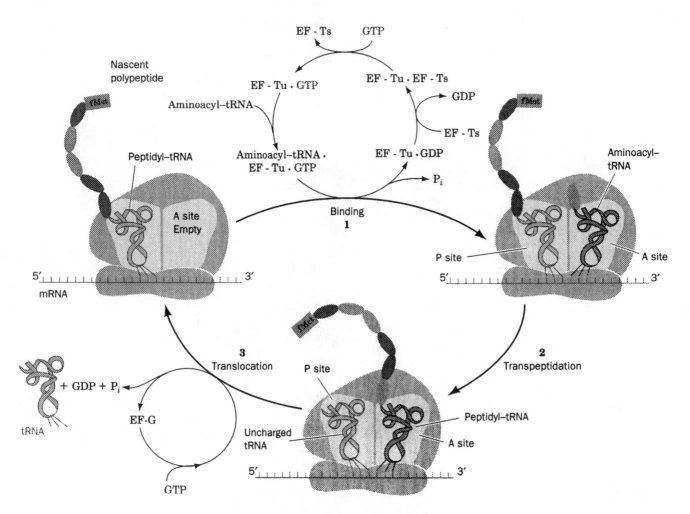

Figure 26-28. *Key to Function.* **The elongation cycle in *E. coli* ribosomes.**

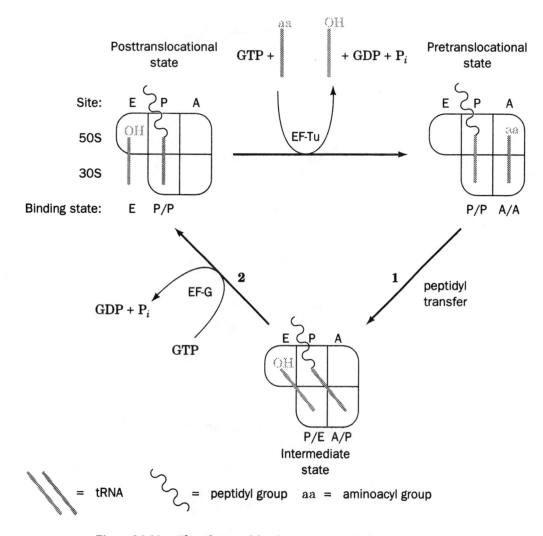

Figure 26-31. The ribosomal binding states in the elongation cycle.

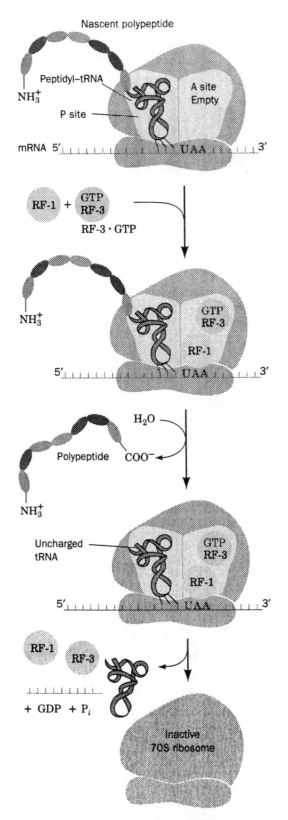

Figure 26-32. The translation termination pathway in *E. coli* ribosomes.

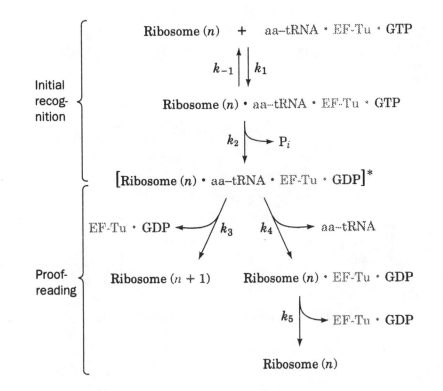

Figure 26-33. A kinetic proofreading mechanism for selecting a correct codon–anticodon interaction.

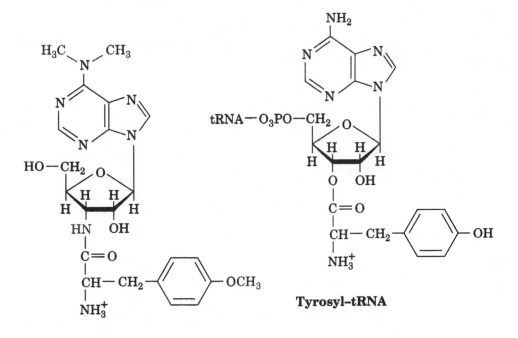

Puromycin

Tyrosyl–tRNA

CHAPTER 27

REGULATION OF GENE EXPRESSION

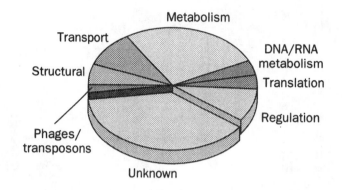

Figure 27-3. Functions of the genes in the *E. coli* genome.

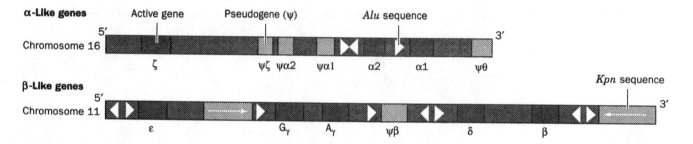

Figure 27-6. The organization of human globin genes.

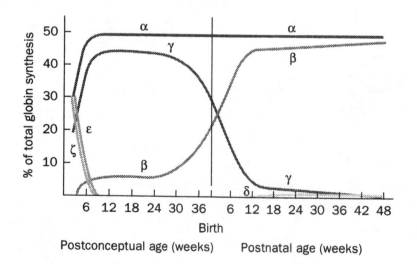

Figure 27-7. The progression of human globin chain synthesis with fetal development.

THE LAC OPERON

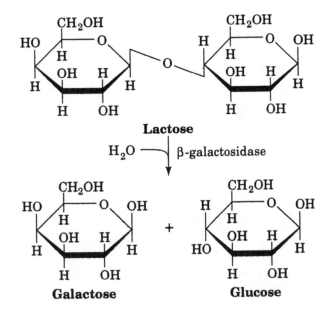

Lactose

H_2O → β-galactosidase

Galactose + **Glucose**

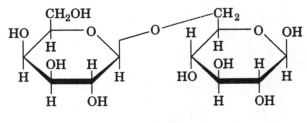

1,6-Allolactose

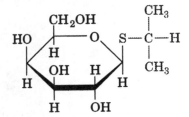

Isopropylthiogalactoside (IPTG)

379

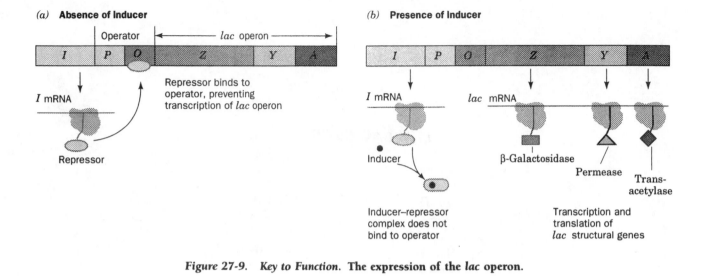

(a) **Absence of Inducer**

(a) **Absence of Inducer**

Operator — *lac* operon —

| *I* | *P* | *O* | *Z* | *Y* | *A* |

I mRNA

Repressor binds to
operator, preventing
transcription of *lac* operon

Repressor

(b) **Presence of Inducer**

| *I* | *P* | *O* | *Z* | *Y* | *A* |

I mRNA   *lac* mRNA

Inducer

β-Galactosidase

Permease

Trans-
acetylase

Inducer–repressor
complex does not
bind to operator

Transcription and
translation of
lac structural genes

Figure 27-9. *Key to Function.* **The expression of the *lac* operon.**

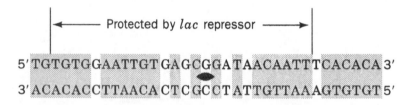

Protected by *lac* repressor

5′ TGTGTGGAATTGTGAGCGGATAACAATTTCACACA 3′
3′ ACACACCTTAACACTCGCCTATTGTTAAAGTGTGT 5′

Figure 27-10. **The base sequence of the *lac* operator O_1. Its symmetry-related regions, which comprise 28 of its 35 bp,
are shaded in red.**

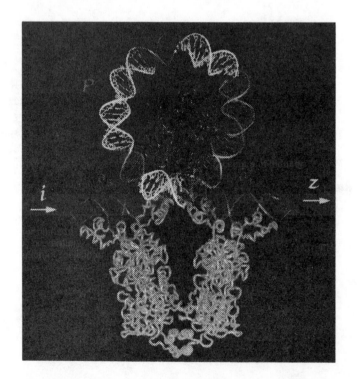

Figure 27-13. **A model of the 93-bp loop formed when the *lac* repressor tetramer binds to O_1 and O_3.**

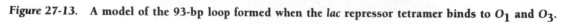

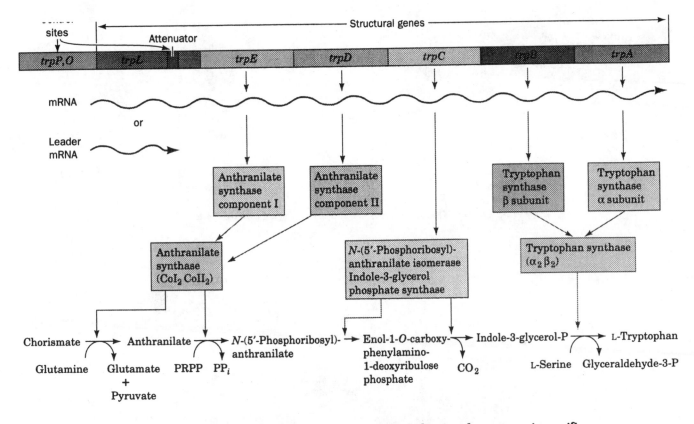

Figure 27-15. A genetic map of the *E. coli trp* operon indicating the enzymes it specifies and the reactions they catalyze.

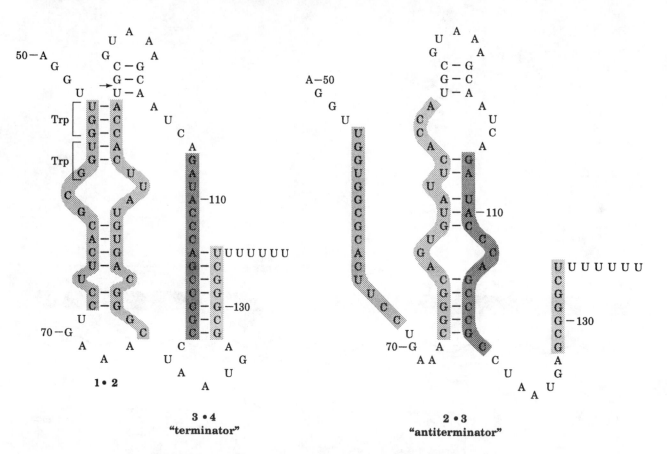

Figure 27-16. The alternative secondary structures of *trpL* mRNA.

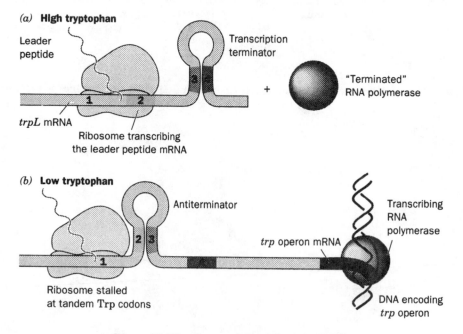

Figure 27-17. Attenuation in the *trp* operon.

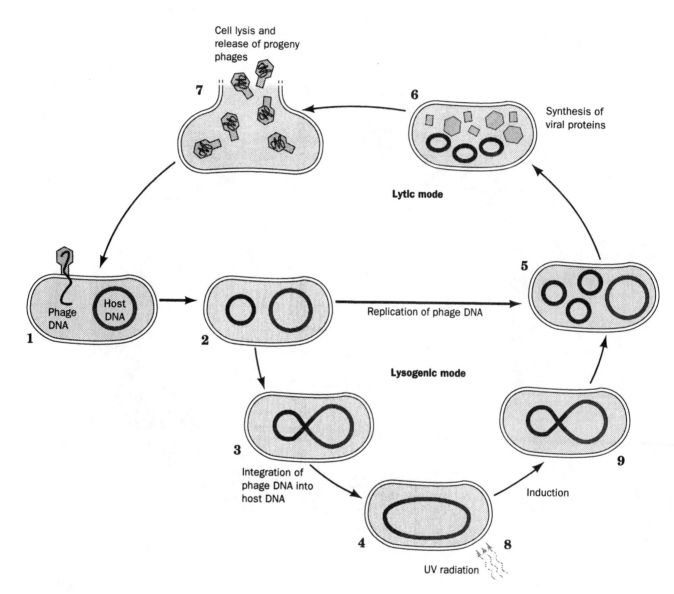

Cell lysis and
release of progeny
phages

7

6

Synthesis of
viral proteins

Lytic mode

Phage
DNA

Host
DNA

1

2

5

Replication of phage DNA

Lysogenic mode

3

Integration of
phage DNA into
host DNA

4

8

UV radiation

9

Induction

Figure 27-18. The λ phage life cycle.

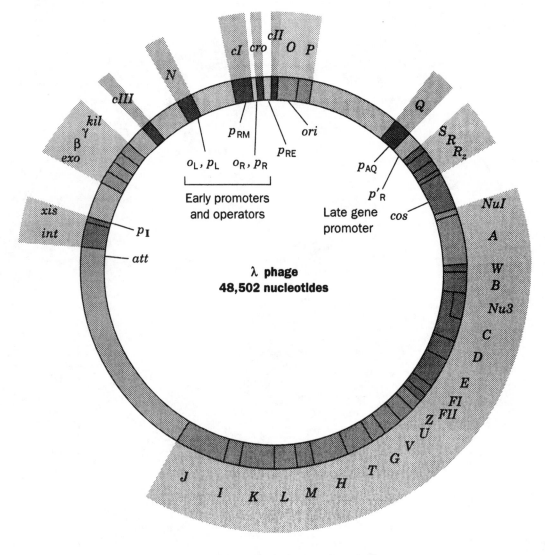

Figure 27-19. A genetic map of bacteriophage λ.

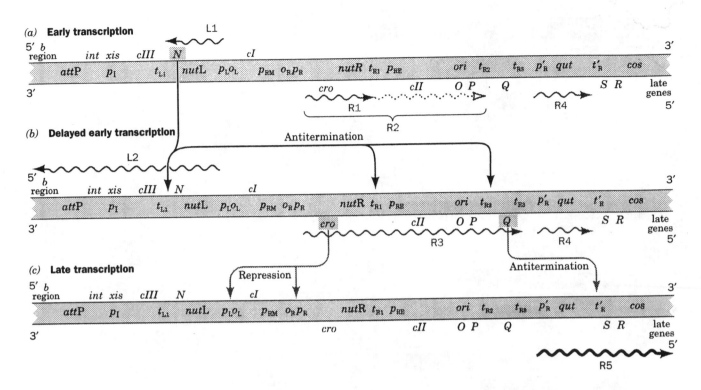

Figure 27-20. Key to Function. Gene expression in the lytic pathway of phage λ.

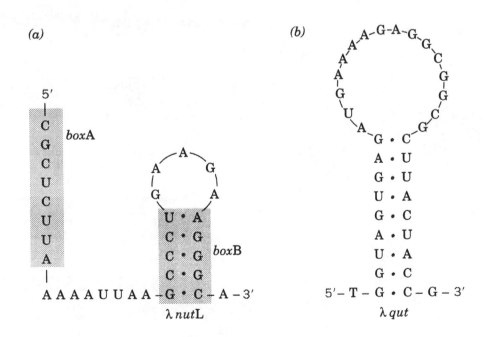

Figure 27-21. The RNA sequences of phage λ control sites.

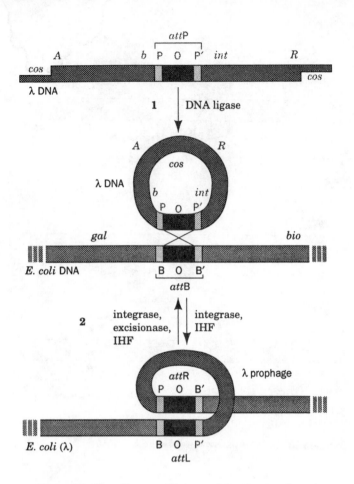

Figure 27-22. Site-specific recombination in phage λ.

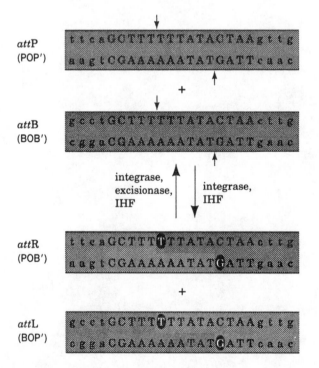

Figure 27-23. The site-specific recombination process that inserts/excises phage λ DNA into/from the *E. coli* chromosome.

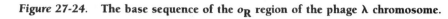

5′ ... TACGTTAAATCTATCACCGCAAGGGATAAATATCTAACACCGTGCGTGTTGACTATTTTACCTCTGGCGGTGATAATGGTTGCA...3′
3′ ... ATGCAATTTAGATAGTGGCGTTCCCTATTTATAGATTGTGGCACGCACAACTGATAAAATGGAGACCGCCACTATTACCAACGT...5′

*p*RM   *o*R3   *o*R2   *o*R1

Figure 27-24. The base sequence of the *o*R region of the phage λ chromosome.

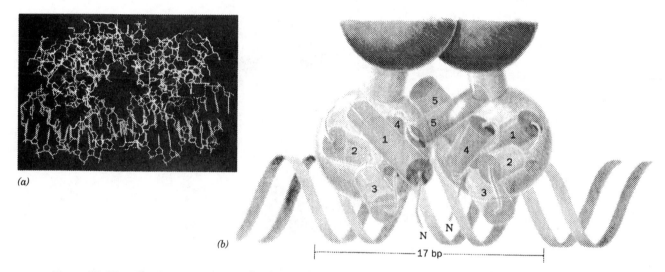

(a)

(b)

|← 17 bp →|

Figure 27-25. The X-ray structure of a dimer of λ repressor N-terminal domains in complex with B-DNA.

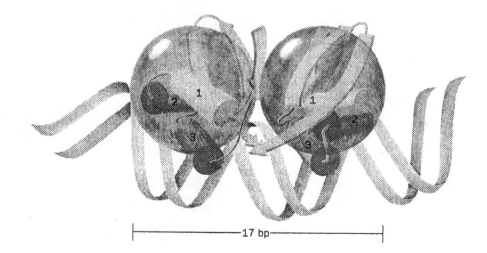

|← 17 bp →|

Figure 27-26. The X-ray structure of the Cro protein dimer in complex with B-DNA.

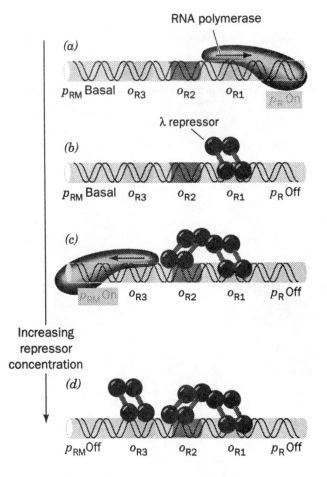

Figure 27-27. The binding of λ repressor to the three subsites of o_R.

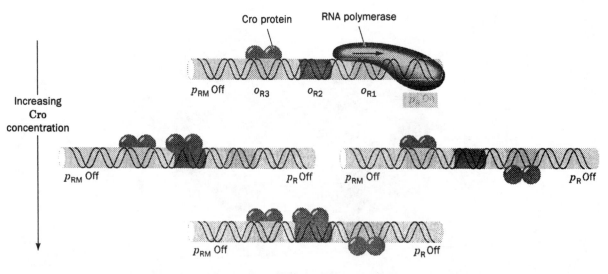

Figure 27-28. The binding of Cro protein to the three o_R subsites.

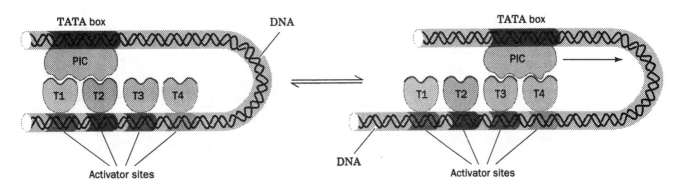

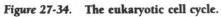

Figure 27-32. A model for the action of transcription factors.

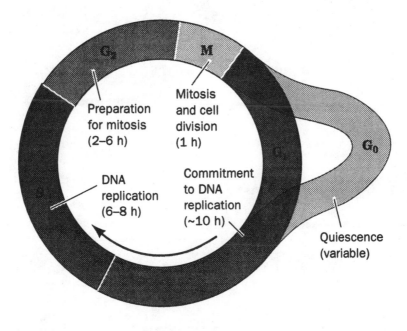

Figure 27-34. The eukaryotic cell cycle.

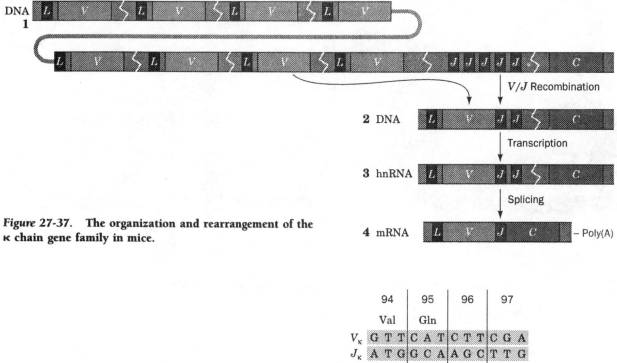

Figure 27-37. The organization and rearrangement of the κ chain gene family in mice.

Figure 27-38. Variation at the V_κ/J_k joint.

	94	95	96	97
	Val	Gln		
V_κ	G T T	C A T	C T T	C G A
J_κ	A T G	G C A	A G C	T T G
			Ser	Leu
	Val	His		
V_κ	G T T	C A T	C T T	C G A
J_κ	A T G	G C A	A G C	T T G
			Ser	Leu
	Val	His		
V_κ	G T T	C A T	C T T	C G A
J_κ	A T G	G C A	A G C	T T G
			Arg	Leu
	Val	His		
V_κ	G T T	C A T	C T T	C G A
J_κ	A T G	G C A	A G C	T T G
			Leu	Leu

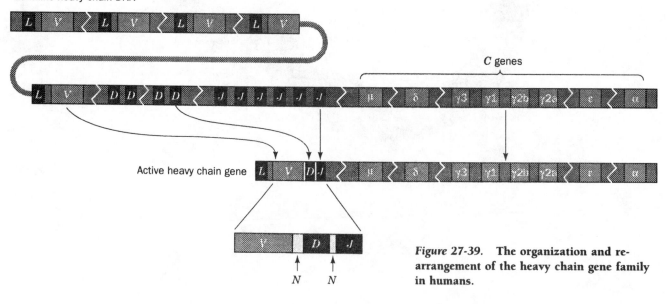

Figure 27-39. The organization and rearrangement of the heavy chain gene family in humans.

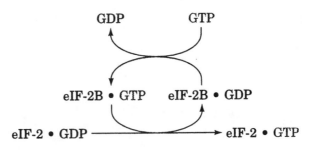

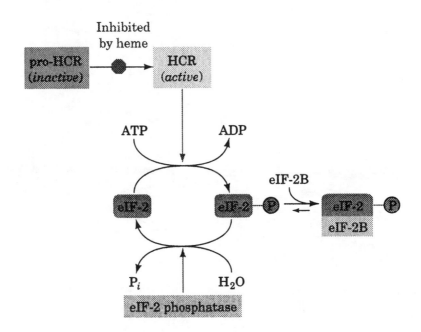

Figure 27-42. A model for heme-controlled protein synthesis in reticulocytes.

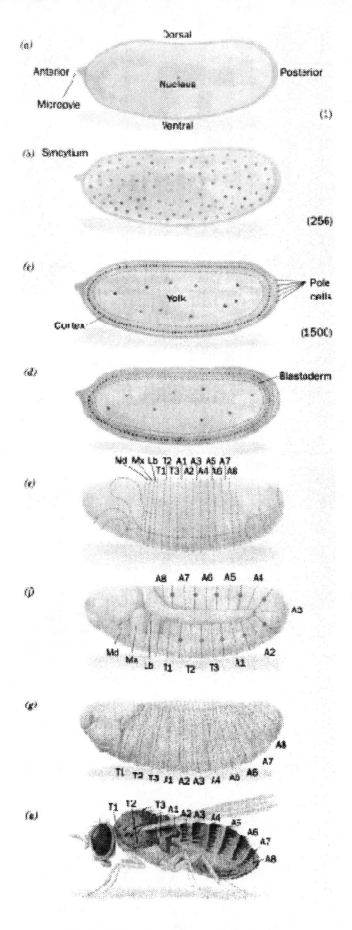

Figure 27-43. Development in *Drosophila*.

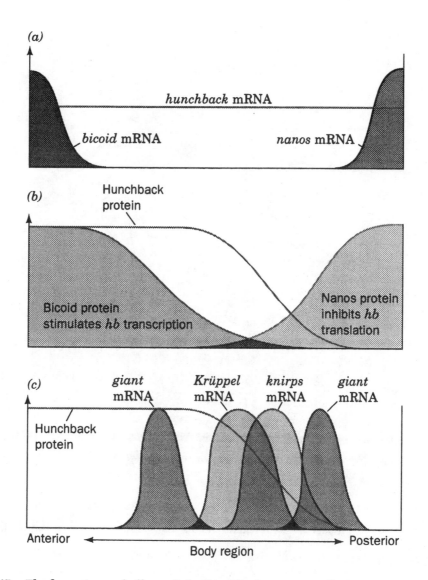

Figure 27-45. The formation and effects of the Hunchback protein gradient in *Drosophila* embryos.

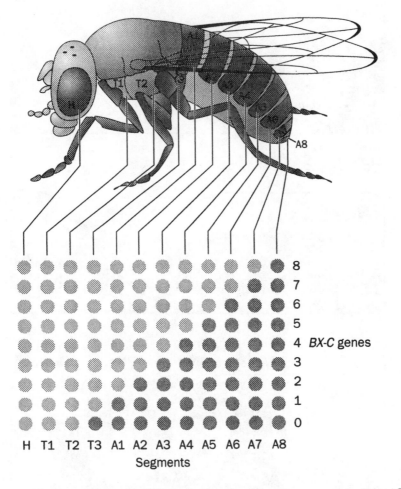

Figure 27-48. A model for the differentiation of embryonic segments in *Drosophila*.